MATHEMATICS FOR STATISTICS

MATHEMATICS FOR STATISTICS

W. L. Bashaw

University of Georgia

John Wiley & Sons

New York · London · Sydney · Toronto

Library of Congress Catalog Card Number: 69-16123
Cloth: SBN 471 05530 1
Paper: SBN 471 05531 X
Printed in the United States of America

To Joan

Preface

THIS book was written to provide the arithmetic and mathematics background that is essential to the learning of introductory applied statistics. The book is addressed to the student who needs to study statistical procedures but who feels insecure about his mathematics preparation.

It is expected that the readers of this book will be highly motivated to learn statistics but do not wish to be forced to enroll in undergraduate mathematics classes for review purposes. Such courses typically minimize applications and include many topics not directly relevant to statistics. The focus of the text is directly on statistical applications. Its relevancy to later statistical studies is kept constantly before the reader. Hopefully, the reading of this book will allow the reader to save much time and money that might otherwise be spent in attending review classes, hiring statistics tutors, and repeating statistics courses.

This textbook was also written to assist the statistics teacher. Many of the problems in the teaching of basic statistics arise from student heterogeneity with respect to mathematical background. The members of typical introductory statistics classes range from students with mathematics majors to those who terminated their training with high school algebra. The typical student has had no more than freshman mathematics. Moreover, students are quite diverse with respect to the number of years that have passed since their last mathematics course.

The teacher usually makes one of two choices. He can refuse to admit students with weak backgrounds, or he can admit all students and accept the accompanying instructional problems. Most teachers make the latter choice for lack of better alternatives.

Hopefully, this book provides an alternative. The teacher can recommend self-study as a prerequisite. Perhaps many schools will be encouraged to offer credit or no-credit classes in basic mathematics for statistics students.

It is common for the teacher to spend considerable time on mathematical notation definitions. He also usually spends countless unplanned hours in and out of the classroom reviewing elementary algebraic manipulations and clarifying elementary mathematical concepts. This time can be better spent directly on statistics topics if all the students have a fairly good mathematics preparation.

It is not rare to find that the *majority* of the students in a first course in statistics have very weak backgrounds. These students find it necessary to struggle through numeric and algebraic manipulations, often preventing them from focusing on the statistical concept that is being taught. Moreover, the inability to perform algebraic operations increases the memory component of learning. Instead of trying to understand how several special-case formulas relate to each other, these students frequently attempt to memorize the entire set of formulas. Thus, a knowledge of algebra is seen to help the student use his study time in learning basic concepts rather than in memorizing. The algebra-wise student has no need for extensive memorization because he can readily derive the various special-case formulas from the general forms.

The problem of student heterogeneity has to be faced by the statistics textbook author. On one extreme, authors assume that students will get the prerequisite training. For example, Quinn McNemar (*Psychological Statistics*, N.Y., Wiley, 1949, p. v) states, ". . . I think it is fruitless to try explaining statistics to those who are not prepared to do some thinking in mathematical language." Other authors try to build in some training. For example, Francis Cornell (*Essentials of Educational Statistics*, N.Y., Wiley, 1956, p. vi) began his text with simple arithmetic and algebra reviews ". . . in view of the limited mathematical training of most education students . . . ." He recognized that the student whose arithmetic skills are weak ". . . spends his time on computation and not on learning statistics."

The book might also encourage bright students with weak mathematical training to continue research methodology studies. The need for scholarly researchers is appalling, and this need cannot be met by restricting student recruitment to those with excellent mathematical backgrounds.

The text is divided into several parts to allow the student to choose the areas of greatest weakness for his review and study. The main parts include basic arithmetic, basic algebra, matrix algebra, set algebra, and a "Miscellaneous Skills" section. Each chapter has several common characteristics. An introductory section outlines the chapter content and relates it to common statistical problems. Thus, the reader begins each chapter with an understanding of its importance in future statistical study.

The emphasis of Part One is rapid and accurate arithmetic computation rather than basic theoretical understandings. Hopefully, most students will be able to go through this section quite easily. Its purpose is to remind the students of those basic numerical relationships and operations that possibly have been forgotten through disuse. This part contains only a few topics that are completely new to the majority of the readers.

Part Two emphasizes rapid and accurate algebraic operations. However, basic theoretical understandings are also stressed. Hopefully, students with previous algebra experience will be able to work through this part fairly efficiently. However, it is expected that many readers will be completely unfamiliar with the included topics. The final chapter on notation, subscripts, and summation will probably be the least familiar.

It is expected that Part Three will be completely unfamiliar to most of the readers. An attempt is made to teach the elements of matrix algebra. The discussion will try to show the value of the matrix language in data recording and data manipulation. It is hoped that general student familiarity with matrix algebra will allow statistics teachers to make more use of this valuable notational system.

Part Four presents set algebra and probability topics. A prevailing trend in the teaching of basic applied statistics is the increasing emphasis on the probability theory foundations of statistics. The use of set algebra greatly facilitates the solution to probability problems; however, few beginning students are conversant with set theory language. The student will be taught some set language and will be introduced to set algebra. This introduction can make the teaching of probability concepts considerably easier.

Part Five includes several topics that do not fit into the previous sections but are no less important. Included is a long chapter on graphing procedures. Also included are chapters on tables and table reading, logarithms, irrational numbers, and computational rules.

I would like to acknowledge the helpful comments of Professors Arthur P. Coladarci, Stanford University, Santo F. Camilleri, Michigan State University, and Richard J. Hill, Purdue University, who reviewed the manuscript for the publisher. I am deeply grateful to Mrs. Charles Wilson and Mrs. G. William Donaldson for the excellent typing and proofreading of a difficult, technical manuscript. I am also indebted to Mrs. Robert Sears who assisted in the typing of an early draft. I would like to thank The Chemical Rubber Company, Cleveland, Ohio, and Barnes and Noble, New York, for the use of the tables that appear in the appendices.

I am especially thankful for the assistance and encouragement of my wife, Joan. Her doctoral training made her an especially good critic of the manuscript.

Athens, Georgia, 1968

W. L. BASHAW

Contents

PART THREE
BASIC MATRIX ALGEBRA

MATHEMATICS FOR STATISTICS

Part One

BASIC ARITHMETIC REFRESHER

You should be able to count, add, subtract, multiply, and divide. If you cannot, you need much more preparation than this book can provide and you should go to basic arithmetic texts as a starting place.

In spite of this assumption of a basic arithmetic ability, it is appropriate to begin the mathematics-for-statistics training at an elementary level—the counting numbers. The reason for this is more to introduce important terminology and notation within a relatively simple context than to review elementary arithmetic topics.

However, there is another important reason. The student of statistics who is slow or confused with regard to elementary arithmetic operations is at an extreme disadvantage in following lectures, doing homework, and completing examinations. The student who can whiz through arithmetic calculations can spend more time trying to understand concepts than can the student who slowly struggles with routine calculations. For this reason, this part hopefully can give you some motivation to practice routine calculations so that you improve your speed and accuracy.

Many students have considerable trouble with routine calculations involving fraction, decimal, and exponent notation. Some students are fairly unfamiliar with negative number operations. Hopefully, problems with these topics can be cleared up with this review.

This part will not take up abstract topics but will retain the pragmatic character of the entire text. Only the commonly used number systems will be discussed. These include the integers, common fractions, the decimal system, and the signed numbers (real-number system).

The examples and problems of this part (and the rest of the book) are frequently taken from statistical applications. You will not always recognize the problems as statistical examples because of your unfamiliarity with common statistical formulas. But as we progress to more complex ideas, the examples will have a more apparent relevance to the study of statistics.

1

The Counting Numbers

Introduction to the Integer System

THIS chapter considers the most common numbers—the common *counting numbers*: 0, 1, 2, 3, 4, etc. These numbers are known by many terms, such as integers, whole numbers, and counting numbers. The term "counting numbers" is probably the most informative. Even in statistics, the use of these numbers is common. We might count the number of subjects in a study, the number of persons voting Democrat, the number of respondents to a questionnaire, or the number of correct answers on an examination.

The arithmetic of integers provides an introduction to many terms and special symbols. The teaching of terminology and notation is the major goal of this chapter; however, you are urged to examine yourself with regard to speed and accuracy of routine calculations.

It will help to begin by distinguishing the symbol for a number from the number itself. Consider the first number of the integer system—the number "one." Many symbols can be used to represent this number. These include, for example,

$$1, \text{one}, \text{I}, \text{ein}, 5 - 4, 5 \times 6 \div 30$$

"One" can be defined (and is defined) quite abstractly by the mathematician. Let's be satisfied to say that "one" means "the characteristic of unity or oneness." That is, a *number* is an abstract concept of numerousness.

On the other hand, the symbols that we read and write are called *numerals*. The numeral for the number one is "I" in Roman numeral notation or "1" in the Hindu-Arabic numeral notation. Other languages have other numerals for the number one. However, we shall use the common Hindu-Arabic notation.

It is important to realize that any set of numerals that means "one" is equivalent and interchangeable with all the others. Thus, "1" and "99 − 98"

3

and "$\frac{1}{2} + \frac{1}{2}$" are all synonomous and interchangeable. Each symbol has the same referent—the number "one."

A very important characteristic of our integer system is that it contains only ten basic symbols—0, 1, 2, 3, 4, 5, 6, 7, 8, and 9. All numerals are constructed with these digits. It is a *base-ten* system. Other systems could be used—such as a *binary* system, which has only *two* symbols, or a dozens-system, which requires twelve symbols. But we commonly use the *base-ten* system requiring *ten* symbols and *only ten* symbols.

Numbers larger than nine are represented by numerals of more than one digit. The *location* of a digit gives the interpretation, since each digit has a *place value*. In the *base-ten* system, the place values are units, tens, hundreds, thousands, ten-thousands, etc. Thus, the numeral "2,314" is interpreted as

> 2 thousands
> 3 hundreds
> 1 ten
> 4 units

It is, literally, the sum of $2000 + 300 + 10 + 4$. The *place-value* of the "2" is the thousands-place; the "3" is in the hundreds-place; the "1" is in the tens-place; and the "4" is in the units-place. The location of the digit in a numeral determines its value in interpreting the numeral.

PROBLEM 1.1

Be sure you understand these terms (operationally, if not formally).

a. Numeral.
b. Number.
c. Base-ten system.
d. Place values of numerals.

Fundamental Operations

It is assumed that you already can add, subtract, multiply, and divide. You are reminded to practice these operations for speed and accuracy. Some problems are presented for a little practice, but you should do a great deal more if you know that your own computations are often slow and inaccurate.

This section will not stress calculations, but will provide some standard language and ideas that might make calculations more rapid and more understandable.

Mathematical operations are indicated by special symbols called *operators*. The *addition operator* is the *plus sign* ($+$). The *subtraction operator* is the *minus sign* ($-$). Multiplication and division operations are designated by several operator symbols, the most common of which are "×" for multiplication and "÷" for division.

ADDITION

Addition is indicated by the plus sign (+) operator. Several multiplication "facts" should be known. (You should also know the sums of all pairs of one-digit numbers.)

First, let's review some terminology. Consider the expression

$$5 + 4 + 1 = 10$$

The numerals "5," "4," and "1" are called *addends*. They are the numbers to be added as directed by the + operators. The numeral "10" is the *sum*. It is the *number* that results when *numbers* are added. (One calculates with numbers, but records numerals.)

Now, let's consider two elementary "facts" by looking at examples. The first set of examples is:

$$3 + 4 = 7 \qquad 1 + 2 = 3 \qquad 0 + 8 = 8 \qquad 7 + 2 = 9$$
$$4 + 3 = 7 \qquad 2 + 1 = 3 \qquad 8 + 0 = 8 \qquad 2 + 7 = 9$$

These are illustrations of the *commutative principle of addition*. This principle, or law, is that *the order of addition is irrelevant*. The numerals can be reordered without changing the answer. Consider these examples:

$$6 + 9 + 4 = 6 + 4 + 9 = 10 + 9 = 19$$

and

$$98 + 73 + 2 = 98 + 2 + 73 = 100 + 73 = 173$$

These are applications of the commutative principle in which numerals were reordered for ease of addition.

Let's take a second set of examples. Look at the *differences* in each pair of expressions:

$$4 + 3 + 1 = 7 + 1 = 8$$
$$4 + 3 + 1 = 4 + 4 = 8$$

$$3 + 7 + 2 = 10 + 2 = 12$$
$$3 + 7 + 2 = 3 + 9 = 12$$

$$6 + 1 + 4 = 7 + 4 = 11$$
$$6 + 1 + 4 = 6 + 5 = 11$$

$$7 + 2 + 9 = 9 + 9 = 18$$
$$7 + 2 + 9 = 7 + 11 = 18$$

These examples illustrate the *associative principle of addition* which is a formal statement to the effect that in a sum of several numbers you can add the first pair first, the last pair, the middle pair, or which ever you like.

The *associative and commutative* laws are quite helpful in speeding up calculations, but they also are directly applied to formal *addition checks*. The most common addition check is to re-add the column of numbers in a different order. For example,

Problem: 1   Check: 5
           4   4
         +5   +1
          ──   ──
          10   10

Subtraction

Subtraction is often referred to as the opposite of addition. (More properly, it is the *inverse* of addition.) The subtraction operator is the minus sign $(-)$. The language of subtraction is presented in the example

$$23 - 15 = 8$$

or, in column notation,

$$
\begin{array}{r}
23 \\
-15 \\
\hline
8
\end{array}
$$

The 23 is called the *minuend,* the 15 is called the *subtrahend,* and the 8 is called the *difference,* or *remainder.*

A fundamental and important fact is the *difference plus* the *subtrahend* equals the *minuend.* This fact is the basis for properly checking subtraction. Since $15 + 8 = 23$, the above subtraction is correct.

Now, consider these examples:

$$8 - 4 = 4, \quad \text{but} \quad 4 - 8 \text{ does not equal 4}$$
$$10 - 3 = 7, \quad \text{but} \quad 3 - 10 \text{ does not equal 7}$$
$$10 - 4 - 3 = 6 - 3, \text{ but it does } not \text{ equal } 10 - 1$$
$$17 - 9 - 5 = 8 - 5, \text{ but it does } not \text{ equal } 17 - 4$$

Thus, there is *neither a commutative nor an associative law* for subtraction as there is for addition.

Also, some subtractions are not defined in the integer system. For example, "$5 - 8$" in the integer system is meaningless. It is similar to saying "a child has 5 marbles and loses 8 of them." (The concept of a *negative number* will be discussed later.) There are no negative numbers among the counting numbers.

Checking Subtraction. The common method of checking subtraction is by the addition of the difference and the subtrahend. The sum is the minuend, if the subtraction is correct. Here are some examples.

1. 35
    −15
    ———
     20   Since 20 + 15 = 35, the answer is correct.

2. 751
    −269
    ———
     482   Since 269 + 482 = 751, the answer is correct.

3. 893
    −127
    ———
     776   Since 127 + 776 = 903, the answer is *wrong*.

MULTIPLICATION

Multiplication is a shortcut method of adding in which all of the addends are the same number. "Seven multiplied by three" means "7 + 7 + 7" *or* "3 + 3 + 3 + 3 + 3 + 3 + 3." The numbers that are multiplied are called *factors*, and the result of the multiplication is called the *product*. In the example "seven times three equals 21," the *factors* are seven and three and the *product* is 21.

There are several operators that all mean "multiply." Some of these may be unfamiliar to you. The most commonly known is the "×" sign, but it might surprise you to know how seldom this multiplication symbol is used. Other common multiplication symbols include a dot and the use of parentheses, braces, and brackets. In some cases, there is *no* multiplication sign of any kind and you are still supposed to know to multiply! So the problem of multiplying seven and three can be written in many ways. Here are *some* of the common ones:

$$7 \times 3 = 21 \qquad 7 \cdot 3 = 21$$
$$(7)(3) = 21 \qquad [7][3] = 21$$
$$7(3) = 21 \qquad 7[3] = 21$$

The first three examples above show what are probably the three most common methods of indicating multiplication. Note that in the dot notation, the dot is raised slightly (7 · 3, not 7.3). This prevents the operator from being confused with a decimal point.

There are several important multiplication facts that should be known. First, notice that $7 \times 8 = 8 \times 7$, $3 \times 2 = 2 \times 3$, $4 \times 7 = 7 \times 4$, and that, in general, the order of multiplication does not affect the product. This is an informal statement of the *commutative principle of multiplication*. It can be applied both to checking multiplication and to rearranging factors to speed up multiplication. It is also a very important aid in simplifying algebraic formulas, as we shall see later.

The second fact is that if more than two products are to be multiplied together, the way one groups the factors does not affect the product. This is the *associative principle of multiplication*. Here are some illustrations:

$$2 \times 3 \times 4 = 6 \times 4 = 24$$
$$2 \times 3 \times 4 = 2 \times 12 = 24$$

$$7 \times 9 \times 3 = 63 \times 3 = 189$$
$$7 \times 9 \times 3 = 7 \times 27 = 189$$

$$8 \times 4 \times 6 = 32 \times 6 = 192$$
$$8 \times 4 \times 6 = 8 \times 24 = 192$$

Moreover, both the commutative and associative principles can be used at one time:

$$8 \times 4 \times 6 = 4 \times 8 \times 6 = 4 \times 48 = 192$$

A third principle of multiplication is equally important. This is the principle of the *distribution of multiplication over addition*, which is often called the "*distributive law of multiplication*." The law is applicable if we wish to multiply a number by a *sum of several numbers*. Here are some examples of the *distributive principle*:

$$3(4 + 2 + 1) = 3(7) = 21$$
$$3(4 + 2 + 1) = 12 + 6 + 3 = 21$$

$$5(9 + 7) = 5(16) = 80$$
$$5(9 + 7) = 45 + 35 = 80$$

$$8(9 + 8 + 1) = 8(18) = 144$$
$$8(9 + 8 + 1) = 72 + 64 + 8 = 144$$

The principle states that the answer (product) is no different whether you find the sum first and multiply the sum by the other factor *or* if you first multiply the factor by every addend and add the resulting products. Again, this principle might help speed up some numeric work, but its most important application is in manipulations of algebraic formulas.

DIVISION

Just as subtraction is called the inverse of addition, division is the inverse of multiplication. The product 3×5 is "defined" in terms of addition as $5 + 5 + 5$. One can similarly "define" division in terms of subtraction. The problem $15 \div 5$ can be stated, "How many times can 5 be subtracted from 15?" We have

$$15 - 5 = 10$$
$$10 - 5 = 5$$

and

$$5 - 5 = 0$$

Five was subtracted from 15 three times, so $15 \div 5 = 3$. Similarly, $16 \div 5$ can be handled by the series of subtractions:

$$16 - 5 = 11$$
$$11 - 5 = 6$$

and

$$6 - 5 = 1$$

In this case, we have a *remainder* of one.

The language of division is important. In the division "$15 \div 5 = 3$," the fifteen is called the "*dividend*," the five is called the "*divisor*," and the three is called the "*quotient*." *Remainders* are expressed either as fractions of the divisor or as numerals labeled *remainder* ($16 \div 5 = 3\frac{1}{5}$ *or* 3 with remainder 1). The *long division* method can also show the terminology:

$$
\begin{array}{r}
3 \quad \text{(quotient)} \\
5\text{(divisor)} \overline{)\ 16 \quad \text{(dividend)}} \\
-15 \\
\hline
1 \quad \text{(remainder)}
\end{array}
$$

Like multiplication, there are many ways to write "divide 25 by 5" or "25 divided by 5." The most common operators have already been used ($\div$ and $\overline{)}$). For example, "divide 25 by 5" can be written

$$25 \div 5 \text{ or } 5\overline{)25}$$

Another very common way of indicating division is *fraction notation*. Fraction arithmetic will be discussed in the next chapter, but it must be stressed here that the fraction symbol is a division operator. In fact, it is *usually* the method of indicating division in algebraic formulas. This notation can be done by a slash (/) or by a horizontal line (—). Thus, "$25 \div 5$" can be expressed as

$$25/5 \quad \text{or} \quad \tfrac{25}{5}$$

or parentheses can be inserted in conjunction with the fraction symbol—e.g., (25)/(5). Note that the number on the left of the slash (or above the line) is the *dividend*. The number on the right of the slash (or below the line) is the *divisor*.

Now let's consider the general laws or principles of arithmetic. First, consider the *commutative principle*. Since $9 \div 3$ and $3 \div 9$ are *not* the same number and $10 \div 2$ and $2 \div 10$ are *not* identical, it is apparent that the *commutative law* is *not* applicable to division—the dividend and divisor cannot be interchanged without changing the quotient.

Also, the *associative principle* does not apply in a series of divisions. For example, $54 \div 9 \div 3$ is actually an ambiguous expression, since if you begin by finding $54 \div 9 = 6$ and then find $6 \div 3$, you obtain 2 as the final quotient. But if you start by finding $9 \div 3 = 3$ and then $54 \div 3$, the quotient is 18. Since the two solutions differ, the *associative principle does not hold for division*.

The ambiguity is generally avoided by using parentheses:

$$(54 \div 9) \div 3 = 2$$
$$54 \div (9 \div 3) = 18$$

The use of parentheses in this manner will be discussed in the next part of this chapter.

The third principle that has been discussed is the *distributive law*. Compare these two problems:

$$\frac{10 + 20}{2} = 15 \quad \text{and} \quad \frac{20}{5 + 5} = 2$$

In the first problem, the application of the *distributive principle* yields $(\tfrac{10}{2}) + (\tfrac{20}{2}) = 5 + 10 = 15$, which is correct. The second example gives $(\tfrac{20}{5}) + (\tfrac{20}{5}) = 4 + 4 = 8$, which is *not* correct. Try similar examples. You will find that division *does* distribute over addition if the sum is the dividend. The distributive principle does *not* hold if the sum is the *divisor*.

A fourth property of division is important. Consider these exercises:

$$2 \div 1 = 2 \quad \text{and} \quad 4 \div 2 = 2$$
$$6 \div 2 = 3 \quad \text{and} \quad 12 \div 4 = 3$$
$$8 \div 2 = 4 \quad \text{and} \quad 16 \div 4 = 4$$
$$100 \div 25 = 4 \quad \text{and} \quad 200 \div 50 = 4$$

Do you see the pattern? Each dividend and divisor in the right hand column is *twice* as large as the dividend and divisor in the left hand column, *but* the quotients are equal! You could take other examples:

$$6 \div 2 = 3 \quad \text{and} \quad 60 \div 20 = 3$$

Here the dividend and divisor are both multiplied by 10, and the quotient remains 3. In general, and you can check this claim by other examples, *if the numerator and denominator are both multiplied by the same number, the quotient is unaffected.* This property might seem trivial at this stage, but it takes on extreme importance in the solution of algebraic formulas and, more important, in the manipulation of statistical formulas.

An equivalent property that might be treated as a fifth principle is that *the division of the dividend and divisor by the same number will not affect the quotient.* For example, $55 \div 15$ is the same as $11 \div 3$ and $99 \div 33$ is the same as $9 \div 3$ or $3 \div 1$. These last two principles are based on the fact that any number divided by itself yields one. So, multiplying or dividing both dividend and divisor by the same number is equivalent to multiplying the entire expression by one, which, of course, has no effect at all on the quotient.

Checking Division. The standard method of checking division is to multiply the quotient by the divisor (and add the remainder, if any, to the resulting product). The result will be the dividend, if the division is correct. For example, $6 \div 2 = 3$ is checked by $2 \times 3 = 6$. Also, $756 \div 13 = 58$ with a remainder of 2 can be checked:

$$
\begin{array}{r}
58 \\
\times 13 \\
\hline
174 \\
58 \\
\hline
754 \\
+2 \\
\hline
756 \\
\end{array}
$$

This check should be a routine part of any division exercise.

Order of Operations and Grouping

There are several general rules that are followed to prevent arithmetic expressions from being ambiguous. Here are some ambiguous expressions:

$$5 \cdot 3 - 2 = ? \quad \text{and} \quad 2 + 8 \div 2 = ?$$

In the first problem, if we multiply first and then subtract, we get 13; however, if we first subtract and then multiply, we get 5. The latter example is similar. We can obtain as answers either 5 or 6, depending on the order of operations. Clearly, we cannot use symbol systems leading to such ambiguity.

There are two general ways of making arithmetic expressions unambiguous. One is by using proper grouping symbols, and the second is by using a standard rule of order of operations. This rule is:

Multiplication and division are conducted prior to addition or subtraction.

Applying this rule to the two examples yields the "correct" answers of 13 and 6, respectively. These answers are "correct" *only* because this is the *standard order of operations.*

Here is a more complex example. What is $10 \times 10 + 2 - 1 + 8 \div 2$? According to the rule we simplify by first multiplying and dividing to get $100 + 2 - 1 + 4$. Now the additions and subtractions are performed to yield 105.

Grouping procedures are the other ways to prevent ambiguities (and they also provide some simple exceptions to the general rule stated above). The common grouping symbols are parentheses, ( ); brackets, [ ]; and braces, { }. Less common grouping symbols involve the square root symbol, $\sqrt{}$, and some fraction notation. Both of these will be covered in later chapters. The general rule for grouping symbols is:

The symbol combinations within a pair of parentheses (or braces or brackets) is to be treated as a single numeral.

So, the first two examples could be written as

$$(5 \cdot 3) - 2 = 13 \quad \text{and} \quad 2 + (8 \div 2) = 6$$

However, the first rule giving multiplication and division priority over addition and subtraction makes the use of parentheses unnecessary. If, on the other hand, we wish to direct you to add before multiplying in the first example and to subtract before dividing in the second example, the use of grouping symbols is necessary. We could say

$$5(3 - 2) = 5 \quad \text{and} \quad (2 + 8) \div 2 = 5$$

Notice that *no* multiplication sign is used in the first example. The operation "multiply" is *designated by the parentheses.* Also notice how the second rule is used. It states to treat the expression within the parentheses as *one* numeral, so we have

$$5(3 - 2) = 5 \cdot 1 = 5$$

and

$$(2 + 8) \div 2 = 10 \div 2 = 5$$

Let's look at more examples and try to get used to working with the two rules.

1. $(2 + 3)(5 + 6) = 5 \cdot 11 = 55$
2. $(2 + 3)5 + 6 = 5 \cdot 5 + 6 = 25 + 6 = 31$
3. $2 + 3(5 + 6) = 2 + 3 \cdot 11 = 2 + 33 = 35$
4. $5[2 + 3(5 + 6)] = 5[2 + 3 \cdot 11] = 5[2 + 33] = 5[35] = 175$
5. $2(2 + 2) + (2 + 2)/4 = 2 \cdot 4 + 4/4 = 8 + 1 = 9$
6. $3 \cdot 5 - 2 + 6 \div 3 + 4 = 15 - 2 + 2 + 4 = 19$
7. $3(5 - 2) + 6 \div (2 + 1) = 3 \cdot 3 + 6 \div 3 = 9 + 2 = 11$
8. $10\{5[(3 + 1)(3 + 2)] + 10(6 - 4)\} = 10\{5[4 \cdot 5] + 10 \cdot 2\}$
   $= 10\{5 \cdot 20 + 20\} = 10\{100 + 20\} = 10\{120\} = 1200$
9. $6[(100 - 50) \div 10] - 20 = 6[50 \div 10] - 20 = 6 \cdot 5 - 20$
   $= 30 - 20 = 10$
10. $[(70 - 25)/5][(50 - 10)/10] = [45/5][40/10] = [9][4] = 36$

Problems for Review

1. Be sure that you know the meaning of the following terms so that you will understand them when your statistics teacher uses them in class.

a. Integer.
b. Numeral.
c. Base 10 number system.
d. Operator.
e. Commutative principle.
f. Associative principle.
g. Distributive principle.

2. Determine which of the following expressions are *true*.

a. $10(6 + 5) = 60 + 50$
b. $100 \cdot 6 + 7 \cdot 4 = 28 + 100 \cdot 6$
c. $5(6 + 5) = 35$
d. $10(5 \cdot 6 + 5) = 350$
e. $(7/7)(8/4)/2 = 1$
f. $10(50 + 1) = 500 + 1$
g. $10[10 + 10]10 = 2000$

h. $(5 + 4 - 5) = 4 + 5 - 5$

i. $8 + 2/2 = 5$

j. $[6 + (4 + 2)]/6 = 1 + 1$

k. $[(2 + 1)(3 + 1)] = 9 + 1$

l. $\{[6 + 4/2] + [7 + 3]\} = 5 + 10$

3. Determine the *integer* that means the same thing as each of these expressions.

a. $7(7 + 7/7)$

b. $[(20 - 5)/3] - [(15 - 12)/3]$

c. $(3 + 5 + 6 + 2)/4$

d. $3 - [(3 + 1 + 2)/3]$

e. $\{6 - [(6 + 2 + 4)/3]\}\{6 - [(6 + 2 + 4)/3]\}$

f. $15 \cdot 22 - 12 \cdot 13 + 6 \cdot 14$

g. $2[3 \cdot 4 + 7 \cdot 8] + 4[3 \cdot 2 + 6 \cdot 3]$

h. $(100 - 75)/(25/5)$

i. $(36/9) + (48/4) + 6(36/9)(48/4)$

j. $(7 \cdot 5 + 3) \div (9 \cdot 2 + 1)$

k. $[(15 - 7)(15 - 7)] \div 32$

l. $12 \times 7 \div 3 - 6 \times 7 \div 21 + 4 \div 2$

m. $[36(7 - 3) + 18(7 - 5) + 5(6 - 2)][4 + 2]$

n. $[6 - (6 + 2)/2][6 - (6 + 2)/2] + [(6 + 2)/2 - 2][(6 + 2)/2 - 2]$

4. Be sure that you can perform *and check* all of the basic operations on integers—addition, subtraction, multiplication, and division.

Common Fractions

Introduction

THE major purpose of this chapter is to refresh your memory of the arithmetic of common fractions and to provide some practice in working with fractions. I was always surprised to see the number of students who had forgotten the basic arithmetic of fractions. Many students are confused about the shortcut "canceling" methods that do much to speed up routine calculations.

This chapter will continue to emphasize the material of Chapter 1 and will strengthen your experience with the grouping and order-of-operation rules. At this point each student should gain sufficient practice with fractions so that working with them becomes routine. Each of the four basic operations—addition, subtraction, multiplication, and division—should be thoroughly mastered with regard to common fractions before proceeding into decimal-fraction arithmetic.

Basic Definitions and Terminology

The concept "fraction" should be considered as part of the extension of the number system from the elementary integer, or counting, system. In this sense, a fraction is merely a *number* that be an integer or a number other than an integer. Some people commonly think of the word "fraction" as meaning a number greater than zero, but less than one. The term "*mixed number*" is commonly used to mean the sum of a fraction that is less than one and a "whole number" (integer).

It is mathematically best to use the word "fraction" more broadly, i.e., the term should include integers, numbers between zero and one, and mixed numbers. In this usage, we include all positive real numbers, or the *positive real number system.*

The term "common fraction" has a less broad meaning; it refers to a particular type of *numeral* for symbolizing the fractional numbers. This *numeral* is the expression of a fraction as a ratio utilizing *divisional notation.* For example, the number one-half can be expressed in common fraction notation as *one divided by two* ("$\frac{1}{2}$") or as *two divided by four* ("$\frac{2}{4}$"). Of course, there are many other ways to write a common fraction numeral that represents "one-half," such as "$\frac{3}{6}$," "$\frac{10}{20}$," "$\frac{400}{800}$," etc.

One technical point needs to be mentioned. Not all fractions can be expressed in common fraction notation. There is a set of numbers called "irrational numbers" that are discussed briefly in a later chapter. Irrational numbers cannot be expressed in either the common fraction nor decimal fraction notation. Decimal notation will be discussed in the next section. Numbers written as decimal fractions can always be written as common fractions, but *not* all numbers that can be expressed as common fractions can be written precisely in decimal notation.

The two parts of a common fraction have technical names. The top number is the *numerator*, and the bottom number is the *denominator*. The numeral "$\frac{3}{4}$" can be used as an example:

$$\frac{\text{numerator} \quad 3}{\text{denominator} \quad 4}$$

The base number, the *denominator*, can be thought of as the *denomination* of the fraction. Just as paper money comes in various denominations, fractions do also—halves, quarters, sixths, eighths, tenths, etc. The top number, the *numerator*, can be thought of as *enumerating* the number of parts of a particular denomination, so that the numeral "$\frac{3}{4}$" means "three units of size one-fourth," or, numerically, it means $\frac{1}{4} + \frac{1}{4} + \frac{1}{4}$.

Addition and Subtraction

Addition and subtraction of common fractions often depend on determining *common denominators.* We *cannot* add or subtract unless all terms in the addition and subtraction have the same denominator. The combining of fractions by addition and subtraction poses a problem if the various terms to be added or subtracted do *not* have the same denominator. For example, to add

$$\frac{7}{16} + \frac{3}{8}$$

the proper first step is to convert *one or both* addends to a new numeral such that both have the same denominator. In the example, the simplest solution is the conversion of $\frac{3}{8}$ to $\frac{6}{16}$. The denominator, 16, is now *common to both addends*, or a *common denominator*. The addition then, is the sum

$$\tfrac{7}{16} + \tfrac{6}{16} = \tfrac{13}{16}$$

For this problem there are other common denominators. *Some* of them include 32, 64, 96, and 128. For example, using a denominator of 32, we have

$$\tfrac{14}{32} + \tfrac{12}{32} = \tfrac{26}{32}$$

which is the same number as $\frac{13}{16}$. However, 16 is the *lowest common denominator*.

There is a very easy way to find at least one common denominator—that is, by *multiplying all denominators*. Here are some examples:

Problem	*One common denominator*
$\frac{1}{2} + \frac{1}{4}$	$2 \cdot 4 = 8$
$\frac{7}{8} - \frac{3}{4}$	$8 \cdot 4 = 32$
$\frac{7}{32} - \frac{4}{23}$	$23 \cdot 32 = 736$
$\frac{1}{2} + \frac{1}{4} + \frac{1}{7}$	$2 \cdot 4 \cdot 7 = 56$
$\frac{1}{10} - \frac{7}{16} + \frac{1}{9}$	$10 \cdot 16 \cdot 9 = 1440$

Problems are sometimes made easier if we can identify the lowest common denominator; however, this is not absolutely necessary, since any common denominator can be used.

The conversion of the original numeral to a new one with the new denominator is based on a fundamental rule: if both the numerator and the denominator of a common fraction are multiplied by the *same* number, then the numeral's referent is not changed. For example, all of these numerals refer to the single number one-half.

a. $\dfrac{1}{2}$

b. $\dfrac{2}{4}\left(\dfrac{1 \times 2}{2 \times 2}\right)$

c. $\dfrac{3}{6}\left(\dfrac{1 \times 3}{2 \times 3}\right)$

d. $\dfrac{10}{20}\left(\dfrac{1 \times 10}{2 \times 10}\right)$

e. $\dfrac{18}{36}\left(\dfrac{1 \times 18}{2 \times 18}\right)$

Using this rule we can solve the problem of converting $\frac{3}{8}$ to $\frac{6}{16}$. Multiplying both the numerator and the denominator by 2 gives $\frac{6}{16}$, or the desired solution. However, consider a less obvious problem:

$$\tfrac{1}{7} - \tfrac{2}{81} = ?$$

Using our first rule, we can quickly find *one* common denominator, namely $7 \cdot 81 = 567$. Notice that, since $7 \cdot 81 = 567$, the correct conversions are

$$\frac{1 \times 81}{7 \times 81} = \frac{81}{567} \quad \text{and} \quad \frac{2 \times 7}{81 \times 7} = \frac{14}{567}$$

and the final answer is $\frac{95}{567}$.

Here is another example:

$$\tfrac{1}{2} + \tfrac{2}{9} + \tfrac{1}{7} = ?$$

A common denominator is $2 \times 9 \times 7 = 126$. The solution is

$$\frac{1 \times 9 \times 7}{2 \times 9 \times 7} + \frac{2 \times 2 \times 7}{9 \times 2 \times 7} + \frac{1 \times 2 \times 9}{7 \times 2 \times 9} = \frac{63}{126} + \frac{28}{126} + \frac{18}{126} = \frac{109}{126}$$

PROBLEM 2.1.

a. $\frac{2}{3} + \frac{1}{4} = ?$ e. $\frac{5}{6} - \frac{1}{3} - \frac{1}{12} = ?$

b. $\frac{1}{8} + \frac{3}{4} = ?$ f. $\frac{1}{9} + \frac{1}{4} + \frac{1}{7} = ?$

c. $\frac{4}{9} + \frac{2}{7} - \frac{1}{9} = ?$ g. $\frac{1}{18} + \frac{1}{2} + \frac{1}{9} = ?$

d. $\frac{3}{16} + \frac{1}{10} = ?$ h. $\frac{5}{6} - \frac{1}{3} - \frac{2}{14} = ?$

"Mixed numbers" present no special problems in addition. One merely adds separately the two whole numbers and the two common fraction terms (after converting to a common denominator). Two cases can arise. Here are examples of both:

Case 1. $7\frac{2}{8} + 5\frac{3}{8} = 12\frac{5}{8}$

Case 2. $7\frac{5}{8} + 5\frac{7}{8} = 12\frac{12}{8}$

In Case 1, $\frac{5}{8}$ is less than one, and the problem is solved satisfactorily. In Case 2, the value $\frac{12}{8}$ is greater than one. It is in fact equal to $1\frac{4}{8}$. The final solution should be written as $13\frac{4}{8}$ or $13\frac{1}{2}$. This solution is no more accurate than $12\frac{12}{8}$, technically speaking; however, it is conventional to convert fractions exceeding one as we have done in Case 2.

Subtraction of mixed numbers also can be presented in two cases. For example,

Case 1. $15\frac{2}{4} - 10\frac{1}{4} = 5\frac{1}{4}$

Case 2. $14\frac{1}{8} - 8\frac{6}{8} = ?$

Case 1 is solved directly by subtracting one whole number from the other $(15 - 10 = 5)$ and subtracting one common fraction from the other $(\frac{2}{4} - \frac{1}{4} = \frac{1}{4})$ to give the solution. In Case 2, we have a problem—subtracting $\frac{6}{8}$ from $\frac{1}{8}$. This is handled by "*borrowing*" $\frac{8}{8}$ (or one) from 14. We can rewrite "$14\frac{1}{8}$" as "$13\frac{9}{8}$." The problem becomes

$$13\tfrac{9}{8} - 8\tfrac{6}{8} = 5\tfrac{3}{8}$$

Here are some more examples that you should study.

a. $1\frac{1}{2} + 1\frac{1}{2} = 2\frac{2}{2} = 3$

b. $1\frac{6}{16} + 5\frac{7}{16} = 6\frac{13}{16}$

c. $1\frac{8}{12} + 2\frac{7}{12} = 3\frac{15}{12} = 4\frac{3}{12}$

d. $9\frac{4}{5} + 8\frac{3}{5} = 17\frac{7}{5} = 18\frac{2}{5}$

e. $2\frac{2}{8} - 1\frac{1}{8} = 1\frac{1}{8}$

f. $10\frac{3}{5} - 7\frac{1}{5} = 3\frac{2}{5}$

g. $8\frac{1}{5} - 2\frac{3}{5} = 7\frac{6}{5} - 2\frac{3}{5} = 5\frac{3}{5}$

h. $7\frac{4}{12} - 3\frac{9}{12} = 6\frac{16}{12} - 3\frac{9}{12} = 3\frac{7}{12}$

PROBLEMS

2.2.
a. $\begin{array}{r} 6\frac{3}{4} \\ +7\frac{2}{4} \\ \hline \end{array}$

b. $\begin{array}{r} 5\frac{1}{8} \\ 7\frac{3}{16} \\ +2\frac{1}{4} \\ \hline \end{array}$

c. $\begin{array}{r} 8\frac{1}{9} \\ 7\frac{2}{3} \\ +1\frac{3}{6} \\ \hline \end{array}$

d. $7\frac{1}{4} + 6\frac{1}{8} + 4\frac{3}{8} = ?$

e. $18\frac{1}{4} + 9\frac{2}{3} + 17\frac{11}{14} = ?$

2.3.
a. $\begin{array}{r} 15\frac{3}{4} \\ -7\frac{6}{9} \\ \hline \end{array}$

b. $\begin{array}{r} 12\frac{8}{9} \\ -11\frac{9}{10} \\ \hline \end{array}$

c. $\begin{array}{r} 45\frac{1}{16} \\ -32\frac{1}{4} \\ \hline \end{array}$

d. $16\frac{1}{5} - 10\frac{1}{8} - 5\frac{1}{4} = ?$

e. $5\frac{3}{16} - 1\frac{2}{8} - \frac{14}{16} = ?$

Multiplication

Multiplication of common fractions is not complicated by the denominator problem. Two common fractions are multiplied by forming a new fraction whose numerator is the product of the numerators of the two factors and whose denominator is the product of the denominators of the two factors.

Here are some examples.

$$\text{a. } \frac{1}{4} \times \frac{1}{2} = \frac{1 \times 1}{4 \times 2} = \frac{1}{8}$$

$$\text{b. } \frac{1}{9} \times \frac{3}{8} = \frac{1 \times 3}{9 \times 8} = \frac{3}{72}$$

$$\text{c. } \frac{3}{4} \times \frac{2}{9} = \frac{3 \times 2}{4 \times 9} = \frac{6}{36}$$

The meaning of fraction multiplication can be illustrated by Example *a*. Example *a* is $\frac{1}{4} \times \frac{1}{2}$, which can be restated, "What is one-fourth of one-half?"

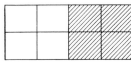

Figure 2.1

We can use a box illustration (Figure 2.1). Let's draw a box and break it into eight equal pieces. Consider one-half of the box (the shaded area). The problem is to find one-fourth of the shaded area. This is *one* of the shaded boxes, or one-eighth of the entire box. That is, one-fourth of one-half is one-eighth!

The multiplication procedure can be explained arithmetically by rewriting Example *c*. Notice that $\frac{3}{4}$ can be written as 3 times $\frac{1}{4}$ and $\frac{2}{9}$ can be written as 2 times $\frac{1}{9}$. Example *c* can be rewritten as

$$3 \cdot \tfrac{1}{4} \cdot 2 \cdot \tfrac{1}{9}$$

This expression can be commuted (reordered) (the commutative principle of multiplication) to give

$$3 \cdot 2 \cdot \tfrac{1}{4} \cdot \tfrac{1}{9} = 6 \cdot \tfrac{1}{36} = \tfrac{6}{36}$$

Mixed numbers present no special problem if you convert all mixed numbers to a form that we will call simple common fractions. For example, the problem $3\frac{1}{4} \cdot 2\frac{1}{8}$ can be solved by standard fraction multiplication if we convert $3\frac{1}{4}$ into $\frac{13}{4}$ and $2\frac{1}{8}$ into $\frac{17}{8}$. The conversions are based on the following schemes:

$$3\tfrac{1}{4} = 3 + \tfrac{1}{4} = \tfrac{12}{4} + \tfrac{1}{4} = \tfrac{13}{4}$$
$$2\tfrac{1}{8} = 2 + \tfrac{1}{8} = \tfrac{16}{8} + \tfrac{1}{8} = \tfrac{17}{8}$$

Of course, you don't need to actually go through all of these steps. Note that

$$3\tfrac{1}{4} = \frac{4 \cdot 3 + 1}{4} = \frac{13}{4}$$

and

$$2\tfrac{1}{8} = \frac{2 \cdot 8 + 1}{8} = \frac{17}{8}$$

and these operations can usually be performed mentally.

In this example, we get $\frac{13}{4} \times \frac{17}{8} = \frac{221}{32}$. It is customary to convert a numeral like this back into mixed fraction form. This is done by dividing 221 by 32. We get $221 \div 32 = 6$ with a remainder of 29, so $\frac{221}{32}$ is equivalent to $6\frac{29}{32}$.

There are other ways to multiply mixed numbers. Frequently, it would be easier to convert all of the numerals to decimal notation and proceed with decimal fraction multiplication.

Canceling. Fraction multiplication frequently can be speeded by *canceling*. This is the process of eliminating certain terms so that small numbers are multiplied instead of larger ones.

Here are some obvious examples.

a. $\dfrac{1}{3} \times \dfrac{3}{9} = \dfrac{1}{\cancel{3}} \times \dfrac{\cancel{3}}{9} = \dfrac{1}{9}$

b. $\dfrac{13}{52} \times \dfrac{12}{13} = \dfrac{\cancel{13}}{52} \times \dfrac{12}{\cancel{13}} = \dfrac{12}{52} \quad \left(\text{instead of } \dfrac{156}{676}\right)$

c. $\dfrac{2}{4} \times \dfrac{4}{16} = \dfrac{2}{\cancel{4}} \times \dfrac{\cancel{4}}{16} = \dfrac{2}{16} \quad \left(\text{instead of } \dfrac{8}{64}\right)$

Canceling usually is done automatically—i.e., if a numerator and a denominator are equal, they are stricken out. However, there is a rationale for doing so based on two facts: multiplication is commutative, and a number divided by itself is one. So we have the following in the three examples.

a. $\dfrac{1}{3} \times \dfrac{3}{9} = \dfrac{3 \cdot 1}{3 \cdot 9} = 1 \cdot \dfrac{1}{9} = \dfrac{1}{9}$

b. $\dfrac{13}{52} \cdot \dfrac{12}{13} = \dfrac{13 \cdot 12}{13 \cdot 52} = 1 \cdot \dfrac{12}{52} = \dfrac{12}{52}$

c. $\dfrac{2}{4} \times \dfrac{4}{16} = \dfrac{4 \times 2}{4 \times 16} = 1 \cdot \dfrac{2}{16} = \dfrac{2}{16}$

Similar to the obvious canceling of identical numerals, one often can perform some elementary divisions by canceling. For example,

$$\tfrac{2}{4} \times \tfrac{1}{8} = \tfrac{1}{2} \times \tfrac{1}{8} = \tfrac{1}{16}$$

Here are some more examples.

a. $\dfrac{7}{8} \times \dfrac{4}{14} = \dfrac{4}{8} \times \dfrac{7}{14} = \dfrac{\cancel{4}}{\cancel{8}_2} \times \dfrac{\cancel{7}}{\cancel{14}_2} = \dfrac{1}{4}$

b. $\dfrac{3}{\cancel{9}} \times \dfrac{\overset{2}{\cancel{18}}}{30} = \dfrac{\cancel{3} \times 2}{\underset{10}{\cancel{30}}} = \dfrac{2}{10}$

c. $\dfrac{\cancel{7}}{8} \times \dfrac{64}{\cancel{7}} = \dfrac{64}{8} = 8$

PROBLEM 2.4.

a. $\frac{1}{2} \times \frac{3}{4} = ?$ g. $\frac{7}{16} \cdot \frac{8}{12} = ?$

b. $(\frac{4}{7})(\frac{3}{8}) = ?$ h. $\frac{6}{9} \cdot \frac{18}{12} \cdot \frac{1}{2} = ?$

c. $(\frac{4}{5})(\frac{5}{4}) = ?$ i. $\frac{6}{19} \cdot \frac{38}{50} \cdot \frac{5}{36} = ?$

d. $1\frac{2}{3} \times \frac{1}{3} = ?$ j. $\frac{4}{5} \cdot \frac{55}{17} \cdot \frac{34}{32} = ?$

e. $(\frac{7}{16})(\frac{3}{10})(\frac{1}{3}) = ?$ k. $\frac{10}{7} \cdot \frac{9}{6} \cdot \frac{8}{5} \cdot \frac{7}{4} = ?$

f. $1\frac{1}{2} \times 2\frac{1}{2} = ?$ l. $3\frac{1}{7} \times 1\frac{33}{44} = ?$

Division

Division of common fractions, like multiplication, is not complicated by the problem of finding common denominators. Also, like multiplication, you can handle mixed numbers satisfactorily by converting to simple common fractions or to decimal fractions. Therefore, we need be concerned only with the problem of dividing one common fraction by another. The traditional rule is to "*invert the divisor and multiply.*" This rule can be applied blindly, although we shall see the rationale behind it.

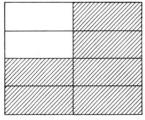

Figure 2.2

Consider several examples of the division procedure according to the rule.

a. $\frac{2}{3} \div \frac{1}{2} = \frac{2}{3} \times \frac{2}{1} = \frac{4}{3} = 1\frac{1}{3}$

b. $\frac{2}{4} \div \frac{2}{3} = \frac{2}{4} \times \frac{3}{2} = \frac{3}{4}$

c. $\frac{3}{7} \div \frac{6}{14} = \frac{3}{7} \times \frac{14}{6} = 1$

d. $\frac{3}{4} \div \frac{1}{8} = \frac{3}{4} \times \frac{8}{1} = 6$

Example *d* can give us insight into the division process. The problem can be stated as the question, "How many eighths are there in three-fourths?" The question can be answered by drawing a figure (Figure 2.2). Consider a square that is divided into eighths. The shaded part represents $\frac{3}{4}$ of the whole square. The division problem is solved by counting the one-eighth-blocks that are shaded. There are 6 such blocks, so

$$\frac{3}{4} \div \frac{1}{8} = 6$$

is correct.

Now, let's look at the arithmetic. We can write the problem as

$$(\tfrac{3}{4}) \div (\tfrac{1}{8}) \quad \text{or} \quad \frac{\tfrac{3}{4}}{\tfrac{1}{8}}$$

Let's multiply the numerator and denominator of this fraction by 8 (or $\tfrac{8}{1}$) to give

$$\frac{(\tfrac{3}{4})(\tfrac{8}{1})}{(\tfrac{1}{8})(\tfrac{8}{1})}$$

which reduces to $(\tfrac{3}{4})(\tfrac{8}{1})$, since $(\tfrac{1}{8})(\tfrac{8}{1}) = 1$.

Consider another example arithmetically:

$$\tfrac{3}{4} \div \tfrac{6}{7}$$

We can write this as

$$\frac{\tfrac{3}{4}}{\tfrac{6}{7}}$$

and then multiplying both numerator and denominator by $\tfrac{7}{6}$, we have

$$\frac{\dfrac{3 \cdot 7}{4 \cdot 6}}{\dfrac{6 \cdot 7}{7 \cdot 6}} = \frac{3 \cdot 7}{4 \cdot 6} = \frac{7}{8}$$

Do you see how these steps lead to the familiar rule, "Invert the divisor and multiply?"

PROBLEMS 2.5.

a. $\tfrac{2}{3} \div \tfrac{1}{6} = ?$ e. $1\tfrac{1}{2} \div \tfrac{3}{4} = ?$

b. $\tfrac{1}{3} \div \tfrac{4}{5} = ?$ f. $3\tfrac{2}{3} \div 6\tfrac{1}{2} = ?$

c. $\tfrac{7}{16} \div \tfrac{8}{16} = ?$ g. $18\tfrac{3}{4} \div 5\tfrac{1}{3} = ?$

d. $\tfrac{9}{12} \div \tfrac{3}{17} = ?$ h. $6\tfrac{1}{2} \div 2\tfrac{3}{4} = ?$

Grouping with Fractions

The fractional division bar is used as a grouping device just as parentheses, braces, and brackets are used to group arithmetic terms. Consider the expression

$$\frac{5 \cdot 11 + 6}{7 \cdot 30 + 4}$$

This is exactly the same as $(5 \cdot 11 + 6) \div (7 \cdot 30 + 4)$. That is, everything above the division bar (the entire numerator) *is to be treated as a single numeral*, just as you do with the expression in parentheses. Similarly, everything below the division bar (the entire denominator) *is treated as a single numeral*. The entire expression simplifies to

$$\frac{55 + 6}{210 + 4} \quad \text{or} \quad \frac{61}{214}$$

Here is another important example:

$$\frac{30 - 5}{5}$$

This is the same as $(30 - 5)/5$ or $(\frac{1}{5})(30 - 5)$. Notice that the following cancellation is *wrong*:

$$\frac{\overset{6}{\cancel{30}} - 5}{\cancel{5}} = 6 - 5 = 1$$

This cancellation violates the principle of treating all of the numerator as a single number. Two correct procedures are as follows.

a. $\dfrac{30 - 5}{5} = \dfrac{\overset{5}{\cancel{25}}}{\cancel{5}} = 5$

b. $\dfrac{30 - 5}{5} = \dfrac{\overset{6}{\cancel{30}} - \overset{1}{\cancel{5}}}{\cancel{5}} = 6 - 1 = 5$

Example *b* is an application of the distributive principle. However, the reader will probably prefer to use method *a*.

Problems for Review

1. Be sure you know the following terms. You will be expected to use them as routine vocabulary terms.

   a. Numerator.
   b. Denominator.
   c. Common denominator.

2. Determine a common denominator for each of these expressions.

   a. $\frac{1}{4} + \frac{1}{6}$ d. $\frac{8}{90} + \frac{6}{30}$

   b. $\frac{2}{17} + \frac{4}{34}$ e. $\frac{1}{4} + \frac{2}{3} + \frac{1}{5}$

   c. $\frac{7}{16} + \frac{4}{5}$ f. $\frac{1}{6} - \frac{2}{15} + \frac{1}{8}$

3. Convert these mixed fractions to simple common fractions.

a. $3\frac{1}{2}$

b. $5\frac{2}{3}$

c. $7\frac{8}{9}$

d. $22\frac{11}{23}$

e. $20\frac{3}{32}$

f. $103\frac{4}{9}$

4. Convert the following fractions to mixed numbers.

a. $\frac{22}{7}$

b. $\frac{45}{16}$

c. $\frac{94}{23}$

d. $\frac{1205}{9}$

5. Calculate the *simplest* equivalent expression for each of the following expressions.

a. $\frac{1}{2} + \frac{3}{4} + \frac{3}{8}$

b. $\frac{3}{16} + 3\frac{1}{8} + 2\frac{1}{2}$

c. $3\frac{7}{8} + 7\frac{2}{15}$

d. $7\frac{2}{3} - 5\frac{1}{4}$

e. $9\frac{3}{8} - 4\frac{6}{9}$

f. $10\frac{1}{3} - 4\frac{3}{4} - 2\frac{9}{16}$

g. $\frac{3}{4} \cdot \frac{16}{9}$

h. $\frac{7}{8} \cdot \frac{16}{3} \cdot \frac{24}{7}$

i. $\frac{6}{11} \cdot \frac{18}{24} \cdot \frac{2}{3}$

j. $3\frac{2}{3} \times 6\frac{3}{5}$

k. $16\frac{2}{9} \times 20\frac{1}{4}$

l. $\frac{3}{4} \div \frac{6}{9}$

m. $\frac{7}{5} \div \frac{2}{3}$

n. $3\frac{2}{3} \div \frac{11}{24}$

o. $6\frac{1}{8} \div 2\frac{1}{16}$

p. $\dfrac{6\frac{1}{8}}{3\frac{1}{16}}$

q. $\frac{1}{5}(6\frac{1}{4} - 3\frac{1}{2})$

r. $(6\frac{2}{3} - 1\frac{2}{6})(4\frac{2}{3} + 2\frac{1}{2})$

s. $(6\frac{1}{4} - 3\frac{1}{8}) \div [(3\frac{1}{2} - 2)/4]$

t. $(7\frac{1}{4} + 2\frac{1}{8})(4\frac{2}{3} - 1\frac{3}{5})/(6\frac{1}{4} - 2\frac{1}{5})$

u. $\dfrac{7\frac{3}{4} + 1\frac{1}{2}}{6\frac{2}{3} + 2\frac{3}{4}}$

The Decimal System

Introduction

THE major purpose of this chapter is to remind you of the arithmetic of decimal fractions. The major problem that must be eliminated is the difficulty in locating the decimal point properly.

The decimal numeral system is a notation procedure for expressing real numbers in base-ten notation. In particular, it is the expression of all fractions in terms of denominators that are multiples of ten—tenths, hundredths, thousandths, millionths, etc. However, instead of writing numerals using the common fraction division notation, the decimal notation is used. That is, instead of "$\frac{1}{10}$," we write ".1"; instead of "$\frac{35}{100}$," we write ".35"; instead of "$\frac{213}{1000}$," we write ".213." The decimal notation is much easier to write and usually results in much easier calculations than the fraction notation.

The decimal point is used as a place holder that separates the numeral for a mixed number into a whole number part and a fractional part. In Chapter 1 we wrote the numeral "2,314" as

> 2 thousands
> 3 hundreds
> 1 ten
> 4 ones

Similarly, we can "explain" the decimal numeral "2,314.219" as

> 2 thousands
> 3 hundreds
> 1 ten
> 4 ones
> 2 tenths
> 1 hundredth
> 9 thousandths

It is literally the sum $2000 + 300 + 10 + 4 + .2 + .01 + .009$. The decimal point separates the units-place from the tenths-place.

Fundamental Operations

ADDITION AND SUBTRACTION

Addition and subtraction of decimals provide no problems other than those of addition and subtraction of integers *if* the decimal points of the addends are lined up. Some examples will demonstrate decimal addition and subtraction.

1. $3.1 + 5.12 + 100.03 =$

$$
\begin{array}{r}
3.1 \\
5.12 \\
+100.03 \\
\hline
108.25
\end{array}
$$

2. $.0035 + .021 + 3.41 =$

$$
\begin{array}{r}
.0035 \\
.021 \\
+3.41 \\
\hline
3.4345
\end{array}
$$

3. $.012 - .0009 =$

$$
\begin{array}{r}
.0120 \\
-.0009 \\
\hline
.0111
\end{array}
$$

4. $4.45 - 1.31 =$

$$
\begin{array}{r}
4.45 \\
-1.31 \\
\hline
3.14
\end{array}
$$

MULTIPLICATION

One simple, but necessary, rule will keep the decimal point straight in multiplication. You should perform the multiplication as if there were no decimals, and then locate the decimal point in the product by this rule.

The number of digits to the right of the decimal point in the product is equal to the sum of the number of digits to the right of the decimal point in the factors.

Consider the problem $2.5 \times .1$. This can be solved by the rule to yield .25. This is obtained by first finding $25 \times 1 = 25$. The proper location of the decimal point is before the "2" since "2.5" has one digit to the right of its decimal point and ".1" has one digit to the right of its decimal point.

A more complex example is 2.31×34.1. First, we find the product 231×341, which is 78771. Next, we observe that the decimal point should be located so that three digits are to its right. The answer is 78.771.

A rationale for the multiplication rule will be presented later in this chapter. For the present, just be sure that you can properly locate the decimal point in the multiplication of decimal numerals. Study the next several examples.

1.	2.13	2.	1.05	3.	.01	4.	12.21
	× .014		× 8.1		× .05		× 31
	852		105		.0005		1221
	213		840				3663
	.02982		8.505				378.51

Note especially Examples 1 and 3 to see how zeros are used as place holders.

DIVISION

Division of decimal fractions can be considered in two cases. Case 1 includes problems in which the divisor is a whole number and the dividend is a decimal fraction. For example, $5.75 \div 25 = ?$ The division is carried out most easily in long-division format. The decimal point in the answer should be placed immediately above the decimal point in the dividend. The solution is

$$
\begin{array}{r}
.23 \\
25\overline{)5.75} \\
5\ 0 \\
\hline
75 \\
75 \\
\end{array}
$$

Note that the actual division process is carried out without regard to decimal points. The decimal point is inserted in the answer immediately above the decimal point in the dividend. Here are some more examples.

1.	2.06	2.	.891	3.	.000075
	14)28.91		110)98.012		50).003750
	28		88 0		350
	91		10 01		250
	84		9 90		250
	7		112		
			110		
			2		

Case 2 consists of problems in which the divisor is a decimal fraction. This case is handled in two steps. First, the problem is revised to make the divisor a whole number, and second, the division on the revised problem is carried out as in Case 1. The revision is based on the fundamental principles that any number multiplied by one is unchanged and that any number divided by itself equals one. Consider the problem $6.25 \div .25$. Let's express the division in fractional notation as

$$\frac{62.5}{.25}$$

Now let's multiply the entire fraction by $\frac{100}{100}$, or one. We have

$$\frac{62.5 \times 100}{.25 \times 100} \quad \text{or} \quad \frac{6250}{25}$$

The quotient of $6250 \div 25$ is the same as the quotient of $62.5 \div .25$, namely, 250. Here are some more examples.

1. $37.5 \div 12.5 = \dfrac{37.5}{12.5} \times \dfrac{10}{10} = 125\overline{)375}^{\,3}$

2. $14.75 \div .75 = \dfrac{14.75}{.75} \times \dfrac{100}{100} = 75\overline{)1475.0}^{\,19.6}$
$$\begin{array}{r} 75 \\ \hline 725 \\ 675 \\ \hline 500 \\ 450 \\ \hline 50 \end{array}$$

3. $.075 \div 1.5 = \dfrac{.075}{1.5} \times \dfrac{10}{10} = 15\overline{)\,.75}^{\,.05}$
$$\begin{array}{r} 75 \\ \hline \end{array}$$

The above explanation and examples show the rationale behind the common decimal division procedure which is, "Move the decimal point in the divisor *and* dividend to the right as many digits as there are digits to the right of the decimal point in the divisor." This procedure commonly uses a karat symbol "$_\wedge$" to show the revised decimal point. The previously used example $6.25 \div .25$ is solved traditionally by moving the two decimal points two places to the right before dividing:

$$
\begin{array}{r}
25 \\
.25_\wedge)\overline{6.25}_\wedge \\
50 \\
\hline
1\,25 \\
1\,25 \\
\hline
\end{array}
$$

The three examples become the following.

1.
$$
\begin{array}{r}
3 \\
12.5_\wedge)\overline{37.5}_\wedge \\
37\,5 \\
\hline
\end{array}
$$

2.
$$
\begin{array}{r}
19.6 \\
.75_\wedge)\overline{14.75_\wedge 0} \\
7\,5 \\
\hline
7\,25 \\
6\,75 \\
\hline
50\,0 \\
45\,0 \\
\hline
5\,0 \\
\end{array}
$$

3.
$$
\begin{array}{r}
.05 \\
1.5_\wedge)\overline{.0_\wedge 75} \\
75 \\
\hline
\end{array}
$$

This standard procedure is essentially the revision of the problem into a simpler one—a problem having an integral divisor.

Sometimes the division will not come out even (as in Example 2, immediately above). In this case, the solution can be written with a remainder. However, usually one continues dividing for several places, stops, and "rounds off" the answer. "Rounding" and division accuracy are discussed in a later chapter. For now, carry out the division for a few places after the decimal point and "round up" or "down" depending on whether the remainder is more or less than one-half of the divisor. The answer to Example 2 could be given as 19.7 or 19.67 or 19.667. The division can be continued for as much accuracy as is desired.

PROBLEMS

    3.1. $5.01 + .1015 + 50.103 = ?$

    3.2. $7.162 + 1.3 + 104 + .0001 = ?$

    3.3. $650 + .05 + 3.75 = ?$

    3.4. $75.32 - 23.13 = ?$

    3.5. $100.05 - 9.134 = ?$

    3.6. $74.21 - 36.9 = ?$

    3.7. $21.3 - 8.279 = ?$

    3.8. $(25.1)(37.41) = ?$

    3.9. $(61.13)(.013) = ?$

3.10. $(71.3)(12.1)(1.05) = ?$

3.11. $14.105 \div .13 = ?$

3.12. $2.56 \div 14.1 = ?$

3.13. $.045 \div 5.1 = ?$

3.14. $51.6 \div .005 = ?$

Conversion of Common Fractions to Decimal Fractions

The conversion of a numeral written in common-fraction notation into an equivalent decimal numeral is done by division. Mixed fractions are no particular problem, since only the fractional part of the mixed number is involved in the transformation. Here are some examples.

1. $\frac{1}{2} = 1 \div 2 = .5$
2. $\frac{24}{72} = \frac{1}{3} = 1 \div 3 = .333 \text{ R } 3$
3. $4\frac{3}{4} = 4 + (3 \div 4) = 4 + (.75) = 4.75$
4. $6\frac{2}{5} = 6 + (2 \div 5) = 6 + (.4) = 6.4$

PROBLEMS 3.15

Rewrite these numerals in equivalent decimal notation.

a. $6\frac{2}{3}$ d. $\frac{2}{250}$

b. $4\frac{1}{5}$ e. $2\frac{5}{8}$

c. $6\frac{1}{25}$ f. $14\frac{3}{8}$

Operations with 10's

Frequently, we must multiply or divide a number by a multiple of 10 such as 10, 100, 1000, 10,000, etc., or .1, .01, .001, etc. There is a direct relationship between the number of zeros in the multiple of 10 and the proper location of the decimal point in the answer.

First, notice that these statements are true:

$$50 \times .01 = 50 \div 100$$
$$75 \times .001 = 75 \div 1000$$
$$6 \times .01 = 6 \div 100$$

That is, for the purposes of this section, we can treat multiplication by a decimal multiple of 10 in its division form.

Let's look at some patterns:

$$5 \times 1 = 5$$
$$5 \times 10 = 50$$
$$5 \times 100 = 500$$
$$5 \times 1000 = 5000$$

The answer in each case is merely 5 with as many zeros attached to it as there are zeros in the multiple of 10. This can be considered as *moving the decimal point to the right*, since each numeral has an implied decimal point just after the units position. This principle carries over to decimal multiplication:

$$3.124 \times 1 = 3.124$$
$$3.124 \times 10 = 31.24$$
$$3.124 \times 100 = 312.4$$
$$3.124 \times 1000 = 3,124$$
$$3.124 \times 10,000 = 31,240$$

That is, to *multiply* any number by a multiple of 10, move the decimal point to the *right* as many places as there are zeros in the multiple of 10.

Division is similar. Consider some examples:

$$\begin{array}{r} 6.575 \\ \hline 10\,)65.750 \end{array} \qquad \begin{array}{r} .6575 \\ \hline 100\,)65.7500 \end{array}$$

$$\begin{array}{r} .06575 \\ \hline 1000\,)65.75000 \end{array} \qquad \begin{array}{r} .006575 \\ \hline 10,000\,)65.750000 \end{array}$$

In these cases, the answer is equal to the dividend with the decimal point moved *left* as many places as there are zeros in the divisor. Similarly, to multiply 65.75 by .1, we move the decimal point one place *left* to get 6.575, since this is equivalent to 65.75 ÷ 10.

Rationale for the Multiplication Rule

Let's develop, by examples, a rationale for the rule for locating the decimal point in the multiplication of decimal fractions. According to the rule, the following are true:

$$(.5)(3) = 1.5$$
$$(.5)(.3) = .15$$
$$(.5)(.03) = .015$$
$$(.5)(.003) = .0015$$

We can rewrite each of these problems to use the rule for dividing by multiples of ten.

$$(.5)(3) = (5/10)3 = 15/10$$
$$(.5)(.3) = (5/10)(3/10) = 15/100$$
$$(.5)(.03) = (5/10)(3/100) = 15/1000$$
$$(.5)(.003) = (5/10)(3/1000) = 15/10,000$$

The final solutions, 1.5, .15, .015, and .0015, respectively, are obtained by moving the decimal point to the left an appropriate number of places. Notice that the final location of the decimal point is based on the number of zeros in the divisor. The number of zeros, in turn, is exactly equal to the *sum* of the digits to the right of the decimal in each pair of factors.

Estimation

At this point we have gone far enough to introduce a fundamental principle in solving arithmetic problems. This principle can be called a *plausibility principle*. The principle is that we generally should be able to *estimate* accurately enough to tell if a calculated solution is plausible or implausible. If the calculated solution is implausible then you can be certain that you have made a calculation error. Implausible solutions often arise from mislocated decimal points, so it is appropriate to introduce this topic here.

For example, the answer to the problem 636 × .52 should be close to 300. If you get an answer of about 33 or 3307, you should know that the decimal point got lost, since only answers of about 300 are plausible.

We can usually get reasonable estimates by quickly working out the problem with only the first digit of each numeral, using zeros for all other numerals. Thus, an estimated answer to 636 × .52 is 600 × .50 = 300. We can usually get reasonable estimates mentally by sticking to one digit accuracy.

Here are some more estimates.

1. (.013)(.874) is about .008 or .009.

2. 763.21 × 12.43 is about 7000.

3. 22.42 ÷ .013 is about 2000.

4. (74.5)(21.75)(605.32) is about 840,000.

5. (.012)(.842)(.132) is about .0008.

It is a good rule always to estimate your answers before you do finer calculations. In this way, you can know whether or not to be confident with your results.

Problems for Review

Perform the indicated operations. Estimate the answers to check the plausibility of your calculated answers.

1.	1.032	2.	191.053	3.	762.1
	18.12		72.14		128.3
	+ 7.0104		+ 3.901		+ 614.3

4.	73.89	5.	17.91	6.	104.8
	−51.53		− 9.72		− 63.234

7. $16.104 + 4.32 + .041 = ?$

8. $39.004 + .016 + 400 + 1.3 = ?$

9. $7014.1 + 8423.6 + 981.8 = ?$

10. $69.741 - 28.329 = ?$

11. $.0943 - .0895 = ?$

12. $7.46 - .0053 = ?$

13. $(6.41 - .03) + (7.93 - .14) + (6.18 - .25) = ?$

14. $(.73 + .067 + .04 + 1.01) - (.04 + .31 + .086) = ?$

15. Properly place the decimal point in the solutions to the following problems.

   a. $(.0125)(.5) = 625$

   b. $(.606)(1.01) = 61206$

   c. $(7.13)(8.1) = 57753$

   d. $(.001)(.0001) = 1$

   e. $14.4 \div .12 = 12$

   f. $8.888 \div 1.01 = 88$

   g. $.0075 \div .025 = 3$

   h. $46.3 \div 926 = 5$

   i. $7.14 \times 10,000 = 714$

   j. $.0134 \times 1000 = 134$

   k. $.0134 \div 1000 = 134$

   l. $76.43 \div 10,000 = 7643$

   m. $76.43 \div .01 = 7643$

16. $[(6.14)(6.14) + (3.21)(3.21) + (5.1)(5.1)] \div 3.0 = ?$

17. $(4.1 - 3.75) - (3.75 - 3.2) + (6.3 - 3.75) - (3.75 - 1.4) = ?$

18. $[125(101.5) + 75(111.3)] \div (125 + 75) = $?
19. $(.41)(.6) + (.013)(.9) = $?
20. $.71001/.63 = $?
21. $(.71001)(.141)/(.63)(.282) = $?
22. $(.7384/.071)(100) = $?
23. $(5.06)(400.13)(.011) = $?
24. $(6.1)(4.32)/(6.31) = $?
25. $(4.2)(.01)/.6 + (3.1)(2.1)/.03 = $?

Proportions and Percentages

Introduction

PERHAPS the most common statistics ever calculated and reported are frequency counts, proportions, and percentages. The statistics instructor usually assumes that the student knows these concepts and knows how to use them. However, some students do not fully understand proportions and percentages and cannot convert one into the other. If you are such a student, you have a handicap and should learn these topics thoroughly now to avoid confusion during your statistics training.

Proportions and percentages have the same utility. Their purpose is to clarify the interpretation of certain data—usually frequency counts. For example, suppose a particular city contains 4162 Democrats. This statistic really is not very informative. But it would be highly meaningful if you knew that the same number was one-half of the registered voters. The number one-half can be written as a proportion—$\frac{1}{2}$ or .5—or it can be written as a percentage—50% or 50 percent. Both the percentage and the proportion numerals are interpretable, whereas the raw frequency count is relatively unhelpful.

A second major use is in the calculation of probabilities, a major objective of statistical studies. Empirical probability estimates are often proportions. This topic is considered extensively in a later chapter.

A point of frequent confusion is the relationship between frequencies, proportions, and percentages. You must learn these relationships and be able to convert quickly any of these into any of the others. A special section of this chapter will be devoted to these conversions.

Basic Definitions

INTRODUCTION

This section will define each of these concepts: frequency, ratio, proportion, and percentage. The calculation of each will be explained.

FREQUENCY

The word "frequency" is used in statistics to mean the number of things that have a specific description. We have used one example—the number of persons in a specific city who were registered Democrats. If we polled a city population and asked each person his political affiliation, the number of persons saying "Democrat" is the frequency of occurrence of the response "Democrat." If we obtained the number of responses "Republican," "Socialist," "Independent," and other possible party affiliations, we could get a *frequency distribution* of political party membership for the city. A frequency distribution of political party membership would give the exact number of persons polled who are members of each party.

Here are some more examples of common data often reported in terms of frequencies of occurrence and frequency distributions: the number of traffic accidents each month; the number of persons contracting cancer; the number of unemployed adults; the number of enrollees in each of several training programs; the number of children who earn A's, B's, C's, D's, or F's on their report card; and the number of automobiles sold by various manufacturers.

In summary, a *frequency* is determined by a *count* of persons or things or any events of interest. A *frequency distribution* gives the frequency counts for a set of related events or alternative events.

PROBLEM 4.1

Suppose a group of college freshmen had the following ages:

18	17	17	19	21
18	19	17	18	18
19	18	16	18	16
18	19	24	17	18
18	17	21	18	21

a. What is the frequency for an age of 19? Of 17?
b. What is the total frequency distribution?
c. What is the frequency for 22?

RATIO

"Ratio" is a general mathematical term that will be used often. The word "ratio" refers to a number that is a quotient of two given numbers. The ratio of 3 to 1 is $3 \div 1$, or 3; the ratio of 2 to 6 is $2 \div 6$, or $\frac{1}{3}$; the ratio of 1 to 10 is $1 \div 10$, or .1; and the ratio of 4 to 7 is $4 \div 7$, or $\frac{4}{7}$.

PROBLEMS

Express in decimal notation *and* common fraction notation the following ratios.

4.2. The ratio of 1 to 8.
4.3. The ratio of 4124 to 10,310.
4.4. The ratio of 8 to 1.
4.5. The ratio of 6 to 25.

PROPORTION

In statistics, the word "proportion" usually refers to a particular ratio— the ratio of a frequency to the *sum* of *all* of the frequencies in a frequency distribution. If in our political poll we found 4162 Democrats out of 8324 registered voters, we say that "the proportion of Democrats is .5," which means that "half of the voters are Democrats."

In most uses of "proportion," the term refers to the ratio of a part to a whole. Because of this, proportions are usually fractions between zero and one (or equal to zero or one).

Since proportions are usually the ratio of parts to wholes, the interpretation of all proportions is quite easy. A proportion of zero means that there exists no such part, while a proportion of one means the part comprises the whole. These meanings hold whether we are talking about persons, events, things, or whatever.

Let's take an example. Suppose a teacher grading an exam gives the following marks to 20 children: five receive A's, ten receive B's, four receive C's, and one receives an F. The proportion of papers marked "A" is $\frac{5}{20}$, or $\frac{1}{4}$. The proportion marked "B" is $\frac{1}{2}$. What proportion receives "C," "D," or "F"? The answers are $\frac{1}{5}$, 0, and $\frac{1}{20}$, respectively.

Suppose the teacher has another class of 40 students and in this class gives 25 A's, 10 B's, and 5 C's. The two classes are quite different in size. The teacher gave 10 B's in both classes; however, *relatively* more B's were assigned to class one since the proportion of B's in class one is $\frac{1}{2}$, while the proportion of B's in class two is only $\frac{1}{4}$. Can you see how, using proportions, we can make the two classes comparable despite the fact that one was twice as large as the other? In which class did the teacher assign more A's? The proportion of A's in class one is .25, while the proportion in class two is .625, so relatively more A's were assigned in class two.

PROBLEM 4.6

The first problem in this chapter required you to determine a frequency count. Convert the frequencies to proportions to show the proportion of the group of freshmen who is in each age group. What is the sum of the proportions? (It should be *one*.)

PERCENTAGES

Proportions are often expressed as percentages by moving the decimal point two places to the right. This is essentially multiplying the proportion by 100. A percentage sign (%) is added to denote percentages. The term "percent" means literally " per 100." Percentages are used to express frequencies on a common base of 100 to aid interpretation. Here are some examples.

Proportion	Percentage
.5	50
.125	12.5
1.00	100
.005	.5
.064	6.4
.7342	73.42

There is always some confusion over the terms "percentage" and "percent." It is customary to use the term "percent" only when it is preceded by a numeral. The word "percentage" is never preceded by a numeral in standard usage. Thus we say correctly:

The percentage is large.
Eighty percent voted.
The percentage is 80 percent.
What percentage voted? Eighty percent voted.

In review, proportions and percentages are essentially the same, except for the location of the decimal point. In addition, percentages often refer to ratios greater than one, which is usually, but not necessarily, untrue with proportions. Thus, we commonly see statements such as, "The budget was increased by 200%" to mean that the budget was tripled in size. The budget statement could have been made with ratio language—"The ratio of next year's budget to this year's budget is 3." The word "proportion" would seldom be used in this context.

PROBLEMS

4.7. Continue working with the age-distribution problem. Find the percentage of freshmen who are in each age bracket.

4.8. Consider these two frequency distributions of grades. Convert both into percentage distributions to show for each class the percentage of students who earn each grade. Check to see if the percentages add up to 100%.

Class One			Class Two	
Grade	Frequency		Grade	Frequency
A	5		A	25
B	10		B	10
C	4		C	5
D	0		D	0
F	1		F	0

Conversions of Frequencies, Proportions, and Percentages

In general, there are only two conversion rules to learn. The first is how to calculate a proportion from a frequency (and its converse—how to calculate a frequency from a proportion). This rule is written in both ways.

1. a. Frequency $\div$ total = proportion
    b. Proportion $\times$ total = frequency

The second rule is how to convert a proportion into a percentage (and its converse—how to convert a percentage into a proportion). This rule is also written in both forms.

2. a. Proportion $\times$ 100 = percentage
    b. Percentage $\div$ 100 = proportion

Rule 2a means, "Move the decimal point two places to the right," while Rule 2b says, "Move the decimal point two places to the left."

Let's learn to use the conversion rules by practicing with these problems.

PROBLEM 4.9

Fill in this table showing average (hypothetical) salaries of 20,000 salesmen.

Salary	Frequency	Proportion	Percentage
$24,000	100		
23,000	120		
22,000	60		
21,000	220		
20,000	400		
19,000	100		
18,000	0		
17,000		.011	
16,000		.010	
15,000		.029	
14,000		.035	
13,000		.064	
12,000		.111	
11,000		.1905	
10,000			8.75
9,000			12.2
8,000			8.5
7,000			6.5
6,000			4.2
5,000			6.4
4,000			3.4
Sums	20,000	1.0000	100.00%

Division of Several Numbers by One Number

Frequently there is reason to divide many numbers by the same divisor (as you did for part of the last problem). These divisions can sometimes be expedited by using multiplication instead of division. This trick is especially helpful if desk calculators are available.

Suppose you want to divide each of these six numbers by 50.

1. 35   4. 4
2. 20   5. 17
3. 15   6. 23

This can be done by six divisions. Or, you can observe that

$$35 \div 50 = 35 \cdot \tfrac{1}{50}$$

and $\frac{1}{50} = .02$. So, $35 \div 50$ is equivalent to $(35)(.02)$, or $.7$. The divisions are as follows.

1. $35 \div 50 = (35)(.02) = .7$
2. $20 \div 50 = (20)(.02) = .4$
3. $15 \div 50 = (15)(.02) = .3$
4. $\ \ 4 \div 50 = (4)(.02)\ \ = .08$
5. $17 \div 50 = (17)(.02) = .34$
6. $23 \div 50 = (23)(.02) = .46$

The choice of dividing six times by 50 or multiplying six times by .02 is arbitrary and entirely a matter of whichever is easiest for you. Usually, the multiplication technique will be faster.

Problems for Review

1. Suppose a teacher has given an examination of 12 questions and has recorded the performance of the class of 8 children in the following table. Correct answers are called "c" and incorrect answers are recorded as blanks. For example, Tom passed questions 1, 4, and 5, and missed all of the other questions.

Table 4.1

						Questions						
	1	2	3	4	5	6	7	8	9	10	11	12
1. Tom	c			c	c							
2. John	c	c		c			c	c	c		c	
3. Mary	c	c		c	c			c	c			
4. Sue	c	c	c	c	c			c	c		c	c
5. Larry	c	c		c	c		c	c	c			
6. Jane	c	c	c	c	c		c			c	c	c
7. Bob	c			c				c				
8. Jack	c	c			c			c				

From the table, determine the number, proportion, and percentage of:

a. Correct answers of each child.
b. Correct answers to each question.

2. Analyze the following two-sentence paragraph. Record the frequency with which the following symbols are used: consonants, vowels (including "y"), numerals, and punctuation marks. Convert the frequencies to proportions and percentages of the total number of symbols.

> The subjects were 317 undergraduate psychology students.
> Of this group, 112 were male and 205 were female.

3. If a firm spends 13% of its funds on development and its total budget was $7,132,000, how many dollars were spent on development?

4. What is the percentage yield on an investment of $1450 if the annual dividend paid was $62.50?

5. If a secretary spends an average of 65% of her 44 hour week typing, how many hours does she type in an average week?

6. If, in the United States, there are 51 female adults to every 50 male adults, what is the proportion and percentage of each sex in the adult population?

Exponents and Radicals

Introduction

THIS chapter serves three purposes. One is to familiarize you with the mathematical language and notation of exponents and radicals. The second is to help you use the special notation as shortcuts in certain arithmetic problems. The third purpose is to show some ways of determining square roots.

The need to handle simple exponential problems occurs quite early in statistical training. At least one needs to learn to handle "squaring" and "taking square roots." These topics are needed as early as the third chapter in many statistics texts.

Definitions and Terminology

EXPONENTS

Exponents provide a shorthand method of writing out multiple multiplications. In particular, if a number is to be multiplied by itself one or more times, the exponential notation is quite helpful. Consider the relationship $3 \times 3 \times 3 \times 3 = 81$. This can be rewritten in exponential notation as $3^4 = 81$. Here, the numeral "4" is the *exponent*. The symbols instruct the reader to multiply three by itself four times ($3 \times 3 \times 3 \times 3$). Here are some more examples:

$$2^2 = 2 \cdot 2 = 4$$
$$2^3 = 2 \cdot 2 \cdot 2 = 8$$
$$2^4 = 2 \cdot 2 \cdot 2 \cdot 2 = 16$$
$$4^3 = 4 \cdot 4 \cdot 4 = 64$$
$$7^2 = 7 \cdot 7 = 49$$

The sentence "$3^4 = 81$" is read, "Three raised to the fourth power is 81" or, more simply, "Three to the fourth power is 81." The exponent "2" is usually read as "squared" rather than "raised to the second power." Thus, "$5^2 = 25$" is read, "Five squared equals 25."

PROBLEMS

Determine the number that is equivalent to each of the following expressions.

5.1. Seven to the third power. $7 \cdot 7 \cdot 7$
5.2. Five raised to the fourth power. $5 \cdot 5 \cdot 5 \cdot 5$
5.3. Two to the sixth power. $2 \cdot 2 \cdot 2 \cdot 2 \cdot 2 \cdot 2$
5.4. 4^3 $4 \cdot 4 \cdot 4$
5.5. 7^2 $7 \cdot 7$
5.6. 10^4 $10 \cdot 10 \cdot 10 \cdot 10$
5.7. 16^2 $16 \cdot 16$
5.8. 3^5 $3 \cdot 3 \cdot 3 \cdot 3 \cdot 3$
5.9. One-half squared. $\frac{1}{2}$
5.10. 1.23^2 1.23×1.23

ROOTS

Determining a root is the inverse operation of powering. Thus, the question, "What is the square of five?" calls for powering five, i.e., $5^2 = 5 \cdot 5 = 25$. On the other hand, the question, "What is the *square root* of 25?" calls for determining the number, which, when multiplied by itself, yields a product of 25. The operator for square root determination (or extraction) is the radical sign: $\sqrt{}$. Thus, "$\sqrt{25} = ?$" is read, "What is the square root of 25?" The answer, of course, is five, since $5 \cdot 5 = 25$.

We can also seek higher order roots just as we can use any number as an exponent. The symbols "$\sqrt[3]{125} = ?$" means, "What number when multiplied by itself three times yields 125?" The equation "$\sqrt[6]{64} = ?$" means, "What number when multiplied by itself six times yields 64?" The answers are 5 and 2, respectively, since $5 \cdot 5 \cdot 5 = 125$ and $2 \cdot 2 \cdot 2 \cdot 2 \cdot 2 \cdot 2 = 64$.

Although, in the study of statistics, there are some occasions when high-order roots might be necessary, usually only square roots are needed. For this reason, only the determination of square roots will be covered in this text.

PROBLEMS

Work the following examples by trial and error. Check your answers by multiplication.

5.11. $\sqrt{16}$ 5.14. $\sqrt{225}$

5.12. $\sqrt{121}$ 5.15. $\sqrt{36} + \sqrt{25}$

5.13. $\sqrt{400}$ 5.16. $\sqrt{.16}$

Square Root Tables

The easiest way to obtain squares and square roots is by tables, such as Table A in the appendix. Table A can be used to square numbers and to take square roots.

SQUARING

The squares of numbers from 100 to 999 are given directly. Other values can easily be obtained from the two-page table. In order to do some examples, here is a small part of Table A.

N	0	1	2	3	4	5
10	10000	10201	10404	10609	10816	11025
11	12100	12321	12544	12769	12996	13225
12	14400	14641	14884	15129	15376	15625

The columns give the third significant figure, so $101^2 = 10201$, $102^2 = 10404$, $103^2 = 10609$, $123^2 = 15129$, $125^2 = 15625$, etc.

Other values can be obtained by applying our knowledge of multiplying by multiples of 10. For example, if you need to find 10.2^2 instead of 102^2, note that $10.2 = 102 \div 10$, so $10.2^2 = (102 \div 10)(102 \div 10) = 102^2 \div 100$. But, $102^2 = 10404$ (from Table A), so $10.2^2 = 10404 \div 100$, or 104.04. Let's take another example. Find 1230^2. First, note that $123^2 = 15129$. Second, note that $1230 = 123 \times 10$ and $1230^2 = 123^2 \times 100$. So, $1230^2 = 15129 \times 100 = 1,512,900$.

Now, answers may not be perfectly accurate. Errors arise from rounding table entries. However, the table is accurate enough for most work. Greater accuracy can be obtained from hand multiplications or by using desk calculators or computers. Slide rules are useful, but generally less accurate than the table.

PROBLEMS

Carry out the indicated operations. Be sure to estimate answers so that you can have confidence in your results. Use Table A.

5.17. 287^2	5.20. 61^2	5.23. 951^2
5.18. 912^2	5.21. 75^2	5.24. 16.5^2
5.19. 738^2	5.22. 16^2	5.25. $.0191^2$

SQUARE ROOTS

The table is used more or less backward to find square roots, since this is the inverse operation of squaring. Suppose the problem is "$\sqrt{17689}$ = ?" Look in the body of the table for "17689," then read the answer from the row and column headings. We get $\sqrt{17689} = 133$, because "17689" is in the row labeled "13" and it is in the column labeled "3."

Here are some more solved examples:

$$\sqrt{10609} = 103; \quad \sqrt{196} = 14; \quad \text{and} \quad \sqrt{15129} = 123$$

Suppose we do not want to know $\sqrt{10609}$, but we need to find $\sqrt{106.09}$. We use the same method as before—note that $106.09 = 10609 \div 100$; therefore, $\sqrt{106.09} = \sqrt{10609} \div \sqrt{100} = 103 \div 10 = 10.3$. However, you *must be careful*. Here it is *critical* that you *know beforehand* that the answer is about 10. The necessity for the warning is clear if you consider $\sqrt{10.609}$. Here the answer must be about 3, since $3^2 = 9$, which is close to 10.609. To find $\sqrt{10.609}$ you must look through Table A for "1060.9" or "106090." Neither appears in the table, but the numeral "106276" does, and this is close enough for most purposes. We obtain $\sqrt{10.609} = 3.26$, approximately. Greater accuracy can be obtained by *linear interpolation*, and this is explained in Chapter 21.

Let's take some more examples. You should verify these results yourself.

$\sqrt{1.7161} = 1.31$	$\sqrt{15625} = 125$
$\sqrt{171.61} = 13.1$	$\sqrt{1.5625} = 1.25$
$\sqrt{17.161} = 4.14$	$\sqrt{15.63} = 3.95$
$\sqrt{171610} = 41.4$	$\sqrt{.1563} = .395$

Many numerals will not be in the table ("17.16" and "15.63," for example). For these, choose the closest value, or use a more accurate method such as interpolation or direct calculation by hand, calculator, or computer.

Problems

Work these exercises for practice. Use the closest table figure as an estimate of the answer.

5.26. $\sqrt{9.734}$

5.27. $\sqrt{27.35}$

5.28. $\sqrt{85.93}$

5.29. $\sqrt{8.59}$

5.30. $\sqrt{110}$

5.31. $\sqrt{2735}$

5.32. $\sqrt{6341}$

5.33. $\sqrt{.6341}$

Direct Square Root Calculation

The direct square root calculation is tedious. It is unlikely that committing the procedure to memory is at all profitable. However, it might be useful on occasion, and for this reason it is included here for reference.

Suppose we wish to obtain an accurate estimate of the square root of 198.81. The calculations are similar to those for long division. First, write out the number under the radical sign and break it into groups of two digits, starting at the decimal point.

1. $\sqrt{1\ 98.\ 81}$

Second, estimate the square root of the first group of digits. In the example, the first group is "1" and the square root of 1 is 1. This is the first digit in the answer, so write it above the "1" in the problem.

2. $\sqrt{\overset{\displaystyle 1}{1\ 98.\ 81}}$

Third, square the estimate ($1^2 = 1$) and subtract it as you would in long division. Bring down the next two digits.

3.
$$
\begin{array}{r}
1 \\
\sqrt{\quad 1\ 98.\ 81} \\
-1 \\
\hline
98
\end{array}
$$

Fourth, add a zero to the "1" to make it "10" and *double* this to make it "20." Write the "20" to the left of the "98."

4.
$$
\begin{array}{r}
1 \\
\sqrt{\quad 1\ 98.\ 81} \\
-1 \\
\hline
20 \quad\quad 98
\end{array}
$$

Fifth, the 20 is to be divided into 98, after an adjustment. Since $98 \div 20 = 4$, or more, let "4" be the next digit in the answer. Add the four to the 20 to give 24. Subtract 4×24 from the 98. Bring down the next two digits.

5.
$$
\begin{array}{r}
1\ \ 4 \\
\sqrt{}\ \overline{1\ 98.\ 81} \\
-1 \\
\hline
\end{array}
$$

$$
\begin{array}{r}
20 \qquad 98 \\
+\ 4 \\
\hline
24 \qquad -96 \\
\hline
2\ 81
\end{array}
$$

Sixth, Steps 4 and 5 are repeated. Add a "0" to the "14" to make it "140," and double the 140 to give 280. $281 \div 280$ is about one, so the next digit in the answer is "1." Add the 1 to the 280 and divide the 281 by the result. Subtract 1×281 from 281 to show no remainder. Insert the decimal point. The solution (14.1) is complete, and there is no rounding error in this example.

6.
$$
\begin{array}{r}
1\ \ 4.\ \ 1 \\
\sqrt{}\ \overline{1\ 98.\ 81} \\
-1 \\
\hline
20 \qquad 98 \\
+\ 4 \\
\hline
24 \qquad -96 \\
\hline
280 \qquad 2\ 81 \\
+\ \ 1 \\
\hline
281 \qquad -2\ 81 \\
\hline
0
\end{array}
$$

Here are some more worked examples. Notice that you can continue the process for as much accuracy as you like.

1.
$$
\begin{array}{r}
2\ \ 5 \\
\sqrt{}\ \overline{6\ 25} \\
-4 \quad \leftarrow 2 \times 2 \\
\hline
40 \qquad 2\ 25 \\
+\ 5 \\
\hline
45 \quad -2\ 25 \leftarrow 45 \times 5 \\
\hline
0
\end{array}
$$

$\sqrt{625} = 25$

2.
$$
\begin{array}{r}
.\ 5\ \ 1 \\
\sqrt{}\ \overline{.26\ 01} \\
-\ 25 \quad \leftarrow 5 \times 5 \\
\hline
1\ 01
\end{array}
\qquad \sqrt{.2601} = .51
$$

$$
\begin{array}{r}
100 \\
+\ \ 1 \\
\hline
101
\end{array}
\qquad
\begin{array}{r}
-1\ 01 \leftarrow 101 \times 1 \\
\hline
0
\end{array}
$$

3.
$$
\begin{array}{r}
2\ \ 7.\ 2\ \ 2\ \ 7 \\
\sqrt{}\ \overline{7\ 41.32\ 00\ 00} \\
4 \qquad \leftarrow 2 \times 2 \\
\hline
3\ 41
\end{array}
\qquad \sqrt{741.320000} = 27.227
$$

$$
\begin{array}{r}
40 \\
+\ 7 \\
\hline
47
\end{array}
\qquad
\begin{array}{r}
-3\ 29 \qquad \leftarrow 7 \times 47 \\
\hline
12\ 32
\end{array}
$$

$$
\begin{array}{r}
540 \\
+\ \ 2 \\
\hline
542
\end{array}
\qquad
\begin{array}{r}
-10\ 84 \qquad \leftarrow 2 \times 542 \\
\hline
1\ 48\ 00
\end{array}
$$

$$
\begin{array}{r}
5440 \\
+\ \ \ 2 \\
\hline
5442
\end{array}
\qquad
\begin{array}{r}
-1\ 08\ 84 \quad \leftarrow 2 \times 5442 \\
\hline
39\ 16\ 00
\end{array}
$$

$$
\begin{array}{r}
54440 \\
+\ \ \ \ 7 \\
\hline
54447
\end{array}
\qquad
\begin{array}{r}
-38\ 11\ 29 \leftarrow 7 \times 54447 \\
\hline
1\ 04\ 71
\end{array}
$$

In any of the subtraction steps, an estimate can be made too large or too small and may need to be changed. For example, let's look at Example 2 above. If the square root of 26 had been estimated as 6, this is clearly too large, since you would be trying to subtract 36 from 26:

$$
\begin{array}{r}
.\ 6 \\
\sqrt{}\ \overline{.26\ 01} \\
-\ 36
\end{array}
$$

If the estimate had been 4, then we would have the following:

$$
\begin{array}{r}
.\ 4\ \ 9 \\
\sqrt{}\ \overline{.26\ 01} \\
-\ 16 \\
\hline
10\ 01
\end{array}
$$

$$
\begin{array}{r}
80 \\
+\ 9 \\
\hline
89
\end{array}
\qquad
\begin{array}{r}
-\ 8\ 01
\end{array}
$$

Since 89 goes into 1001 more than 10 times, 4 is seen to be too low as an estimate. Similar estimation errors can be made at any step, and if they occur, a reestimation needs to be made.

PROBLEMS

5.34. $\sqrt{2513.0169}$

5.35. $\sqrt{137789.44}$

5.36. $\sqrt{65.98}$

Radical Sign as a Grouping Operator

The square root operator (or radical sign) can be treated exactly like a set of braces or parentheses for indicating the order of operations. For example, $\sqrt{16 + 9}$ is to be treated as identical to $\sqrt{(16 + 9)}$. The answer that is correct in terms of standard convention is 5 and not 7. The numbers 16 and 9 must be added *before* obtaining the square root—$\sqrt{16 + 9} = \sqrt{25} = 5$, and it does *not* equal $4 + 3$ or 7.

5.37. $\sqrt{6(7) - 2(3)} = ?$ $\sqrt{42-6} = \sqrt{36}$

5.38. $\sqrt{(5 - 2)^2 + (4 - 2)^2 + (8 - 2)^2} = ?$

5.39. $\sqrt{178 - 3^2} + \sqrt{101 + 4.5} = ?$

5.40. $\dfrac{65 - 15}{\sqrt{(60 - 10)(75 - 25)}} = ?$

Multiplication and Division with Exponents

There are several special multiplication or division problems that can be simplified by exponent manipulation. Let's look at these by considering examples.

POWER OF A PRODUCT

a. $[(2)(4)]^2 = [(2)(4)][(2)(4)] = (2^2)(4^2)$

b. $[(7)(3)]^3 = [(7)(3)][(7)(3)][(7)(3)] = (7^3)(3^3)$

c. $(3 \cdot 5 \cdot 2)^2 = (3 \cdot 5 \cdot 2)(3 \cdot 5 \cdot 2) = (3^2)(5^2)(2^2)$

d. $(4 \cdot 2)^3 = (4 \cdot 2)(4 \cdot 2)(4 \cdot 2) = 4^3 2^3$

Can you see the pattern? The rule can be stated as, "A power of a product is the product of the powers."

PRODUCT OF POWERS OF THE SAME NUMERAL

a. $5^2 \cdot 5^3 = (5 \cdot 5)(5 \cdot 5 \cdot 5) = 5^5$

b. $4^2 \cdot 4^4 = (4 \cdot 4)(4 \cdot 4 \cdot 4 \cdot 4) = 4^6$

c. $3^3 \cdot 3^3 = (3 \cdot 3 \cdot 3)(3 \cdot 3 \cdot 3) = 3^6$

d. $2^4 \cdot 2^3 = (2 \cdot 2 \cdot 2 \cdot 2)(2 \cdot 2 \cdot 2) = 2^7$

Do you see the pattern? What is the relationship of the exponents on the left to the exponents on the right? The relationships are a, $2 + 3 = 5$; b, $2 + 4 = 6$; c, $3 + 3 = 6$; and d, $4 + 3 = 7$.

The rule in these examples is to *add* the exponents. Other examples are $3^2 \cdot 3^3 = 3^{2+3} = 3^5$ and $9^5 \cdot 9^7 = 9^{5+7} = 9^{12}$. This rule is applied below.

THE POWER OF A POWER

a. $(2^3)^4 = (2^3)(2^3)(2^3)(2^3) = 2^{12}$

b. $(4^5)^2 = (4^5)(4^5) = 4^{10}$

c. $(6^3)^3 = (6^3)(6^3)(6^3) = 6^9$

d. $(5^2)^4 = (5^2)(5^2)(5^2)(5^2) = 5^8$

The pattern in the exponents should be clear. The relationships are a, $3 \times 4 = 12$; b, $5 \times 2 = 10$; c, $3 \times 3 = 9$; and d, $2 \times 4 = 8$. The rule is to *multiply* the exponents.

DIVISION OF EXPONENTS

The principles above also apply to fractions and division. Keep in mind that a division (say, $4 \div 2$) can be treated as a multiplication ($4 \div 2 = 4 \cdot \frac{1}{2} = 2$). Some examples of the rules are:

a. $(4 \div 3)^2 = (\frac{4}{3})(\frac{4}{3}) = 4^2/3^2 = 4^2 \div 3^2$

b. $(18 \div 4)^3 = 18^3 \div 4^3$

c. $(\frac{1}{3})^2(\frac{1}{3})^3 = (\frac{1}{3})(\frac{1}{3})(\frac{1}{3})(\frac{1}{3})(\frac{1}{3}) = (\frac{1}{3})^5$

d. $(\frac{18}{7})^3(\frac{18}{7})^6 = (\frac{18}{7})^9$

e. $[(\frac{3}{4})^2]^3 = (\frac{3}{4})^2(\frac{3}{4})^2(\frac{3}{4})^2 = (\frac{3}{4})^6 = 3^6/4^6$

f. $[(\frac{7}{8})^4]^3 = 7^{12} \div 8^{12}$

Here are some new examples involving division.

a. $5^4 \div 5^2 = \dfrac{5 \cdot 5 \cdot 5 \cdot 5}{5 \cdot 5} = 5 \cdot 5 = 5^2$

b. $7^5 \div 7^2 = \dfrac{7 \cdot 7 \cdot 7 \cdot 7 \cdot 7}{7 \cdot 7} = 7 \cdot 7 \cdot 7 = 7^3$

c. $3^6 \div 3^4 = \dfrac{3 \cdot 3 \cdot 3 \cdot 3 \cdot 3 \cdot 3}{3 \cdot 3 \cdot 3 \cdot 3} = 3 \cdot 3 = 3^2$

d. $3^9 \div 3^4 = 3^{9-4} = 3^5$

e. $18^4 \div 18^2 = 18^{4-2} = 18^2$

Can you see the pattern in the exponents? The rule is to *subtract* the exponents in these examples. This rule can be extended to cover problems like $6^3 \div 6^5$, as we shall see in the next subsections.

Special Exponents

EXPONENT OF ONE

An exponent of one does *not* mean "multiply a number by itself once," as this will be confused with the exponent "two." It means that there is no powering indicated at all. Usually, then, the exponent of one is unnecessary and is not written. For example $6^1 = 6$, and we usually write "6" instead of "6^1." However, it is sometimes helpful to think of a numeral without an exponent as having an exponent equal to one. For example, in the problem

$$7^5 \div 7^4$$

we use the subtraction rule to get $5 - 4 = 1$; so the answer is "7^1." However, we usually write this answer as "7" and not "7^1."

EXPONENT OF ZERO

The zero exponent is helpful on some occasions. Like the exponent of one, its meaning is a standard convention. Let's look at some more exponential patterns.

$$3^5 = 3 \cdot 3 \cdot 3 \cdot 3 \cdot 3 = 243$$
$$3^4 = 3 \cdot 3 \cdot 3 \cdot 3 \quad = 81 = 243 \div 3$$
$$3^3 = 3 \cdot 3 \cdot 3 \quad\quad = 27 = 81 \ \div 3$$
$$3^2 = 3 \cdot 3 \quad\quad\quad = 9 \ = 27 \ \div 3$$
$$3^1 = 3 \quad\quad\quad\quad = 3 \ = 9 \ \ \div 3$$
$$3^0 = \ ?$$

There are no factors of three belonging in the last line since the zero exponent directs us to use *no* factors. We could say $3^0 = 0$, but this is not consistent with the pattern on the right. The last element, to use the same pattern, is $3 \div 3$, which, of course, is one. For *mathematical consistency*, we *define* 3^0 as 1.

Here are some more examples of why the exponent of zero *always* means "one" no matter what the base number is:

$$7^5 \div 7^5 = 1 \qquad \text{but} \qquad 7^{5-5} = 7^0$$
$$5^4 \div 5^4 = 1 \qquad \text{but} \qquad 5^{4-4} = 5^0$$

NEGATIVE EXPONENTS

Exponents with minus signs also have a standard conventional meaning. In general, they represent *reciprocals*. Let's look at a continuation of the pattern of powers of three presented in the last subsection:

$$3^3 = 3 \cdot 3 \cdot 3 = 81 \div 3 = 27$$
$$3^2 = 3 \cdot 3 \qquad = 27 \div 3 = 9$$
$$3^1 = 3 \qquad\quad = 9 \div 3 = 3$$
$$3^0 = \qquad\qquad = 3 \div 3 = 1$$

Let's continue the pattern using minus signs:

$$3^{-1} = 1 \div 3 = \tfrac{1}{3}$$
$$3^{-2} = (\tfrac{1}{3}) \div 3 = \tfrac{1}{9}$$
$$3^{-3} = (\tfrac{1}{9}) \div 3 = \tfrac{1}{27}$$

or, in another form

$$3^{-1} = \tfrac{1}{3}$$
$$3^{-2} = \tfrac{1}{3^2}$$
$$3^{-3} = \tfrac{1}{3^3}$$
$$3^{-4} = \tfrac{1}{3^4}$$

Consider these examples, also.

$$\text{a. } 7^3 \div 7^5 = \frac{7 \cdot 7 \cdot 7}{7 \cdot 7 \cdot 7 \cdot 7 \cdot 7} = \frac{1}{7 \cdot 7} = \frac{1}{7^2}$$

and

$$7^3 \div 7^5 = 7^{3-5} = 7^{-2}$$

$$\text{b. } 6^2 \div 6^5 = \frac{6 \cdot 6}{6 \cdot 6 \cdot 6 \cdot 6 \cdot 6} = \frac{1}{6^3}$$

and

$$6^2 \div 6^5 = 6^{2-5} = 6^{-3}$$

Problems for Review

Carry out the indicated operations.

1. a. 3.102^2
   b. 41.2^2
   c. 3.01^3
   d. 11.03^2
   e. 1.03^3

2. a. $\sqrt{54.4}$
   b. $\sqrt{5.44}$
   c. $\sqrt{544}$
   d. $\sqrt{1032}$

3. a. $(\sqrt{36} + \sqrt{25})^2$
   b. $\sqrt{6^2 + 5^2}$
   c. $\sqrt{6^2 + 4^2 - (2)(.7)(6)(4)}$
   d. $\sqrt{(24 - 4^2)(37 - 5^2)}$
   e. $\sqrt{98.5 - (25^2)/10}$
   f. $\sqrt{(12 - 10)^2 + (10 - 7)^2 + (13 - 10)^2 + (10 - 8)^2}$

4. Simplify the following, using exponent notation.

   a. $(6^3)(6^4)$
   b. $(5^4)^3$
   c. $(7^4)^3 \div (7)^2(7^3)$
   d. $(6^4)^2 \div (6^5)(6^3)$
   e. $25^4 \div 5^2$
   f. $(7^4)(7^3)(5^3) \div (25)(49)$
   g. $6^3 \div 6^5$
   h. $4^2 \div 2^6$

Negative Numbers

Introduction

I⊤ is difficult to do any statistical calculations without knowing fundamental operations on negative numbers. Negative numbers are numbers "less than zero." At one time, mathematicians used the terms "directed" or "signed" numbers for negative numbers.

The basic operations—addition, subtraction, multiplication, and division—must be so well known that they are second nature to you. As you are being presented with the statistical concept "variance," there is little time to be wondering, "Is a minus times a minus positive or negative?" Be sure that you learn these operations thoroughly.

The Real Number System and the Number Line

The "real number system" is the set of numbers that is used in almost all statistical work. You should already be thoroughly familiar with at least the positive real numbers—these include the counting numbers (integers) and all fractions that can be expressed as ratios or decimal fractions. Also included are numbers called "irrational numbers," which are nonintegers that cannot be expressed as fractions. Examples of these are $\sqrt{2}$ and $\sqrt{3}$. These two numbers cannot be expressed exactly by numerals that are in decimal or ratio notation. However, they are among the real numbers. At the end of this chapter you will see some examples of "nonreal" or "imaginary" numbers, which are not numbers in the real number system.

The real number system includes numbers less than zero. In order to distinguish between these two sets of numbers, we call those exceeding zero "positive" numbers. Numbers less than zero are called "negative" numbers.

Negative numbers are expressed as numerals preceded by a minus sign, while positive numbers are indicated by numerals preceded by either a plus sign or no sign.

Negative numbers are common in everyday use. Certainly, everyone uses negative numbers to discuss the winter temperature. In financial affairs, we can use negative numbers. In fact, an overdrawn bank account can be considered as an account containing a negative amount of money. Let's use banking as an example. Suppose I have $100 in my account and in one day I write three checks in the amounts $50, $30, and $45. My checkbook stub should show:

$$
\begin{array}{lll}
\text{1.} \quad \$100 & \text{2.} \quad \$50 & \text{3.} \quad \$20 \\
\underline{- \quad 50} & \underline{- \quad 30} & \underline{- \quad 45} \\
\quad \$\ 50 & \quad \$20 & \quad ?
\end{array}
$$

The final entry should be "−25," indicating that the account is overdrawn by $25.

A common tool for conceptualizing number systems is by a "number line." Since the number line can also help demonstrate some fundamental operations, let's use it to help us understand negative numbers.

Think of a straight line that has no end—an infinitely long line. Let's mark a spot on the line with a zero. The zero will be the basic reference point. Here is a sketch of part of our number line. We will call the right side

the positive side and mark it with a plus sign. The left side is the negative side and is marked with a minus sign.

Notice that the terms "positive," "negative," "plus," and "minus" are, in a way, quite arbitrary labels that are used to distinguish two sets of numbers or to distinguish the two sides of the number line. We could have used the terms "left" and "right" instead of "negative" and "positive." However, the language is not arbitrary in the sense that we need a common, standard vocabulary, so we will continue to use the customary mathematical terminology.

Let's define an arbitrary unit of length and subdivide the line into sections of this unit length, starting at the zero point and working both ways. At the end of each length we will mark the line.

Let's define our unit as having a length of "one" (or "one unit"—this could be one inch, one yard, one millimeter, or one of any arbitrary unit). Now we label our marks on the line with signed numbers, as in the next diagram.

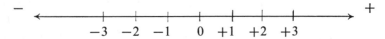

If you think of the line as infinite in length, you can conceptualize any real number as having a corresponding point somewhere on the line. The number "100" corresponds to a point 100 units to the right of 0. The number "−50" corresponds to a point 50 units to the left of 0.

In fact, we can now *define a negative number* as any number corresponding to a point on the number line to the left of zero. This is a perfectly good definition.

Fractions and irrational numbers are also on the line somewhere. For example, the interval from +1 to +2 contains an infinite number of points, each corresponding to fractional numbers. Contained in this interval are points corresponding to $1\frac{1}{2}$, 1.1032, $1\frac{3}{4}$, 1.9995, $\sqrt{2}$, $\sqrt{3.9}$, and any of an infinite number of other numbers that are greater than +1, but less than +2.

PROBLEM 6.1

Sketch a number line; indicate on it the approximate location of these numbers: $-3, 2.5, +1\frac{3}{4}, -\frac{1}{2}, .01$, and $-\sqrt{2}$.

Addition of a Positive Number to a Real Number

Let's learn to add using the number line. The addition operation will be defined as "move to the right on the number line." The problem, "What is 5 + 2?" will be defined as follows. Start at 5 on the number line and move right two units. The answer is the number corresponding to the stopping place. Since "7" is the numeral on the line at the stopping point, 5 + 2 = 7.

Well, that was simple enough. But what about $(-5) + 2 = ?$ Locate "−5" on the number line and move to the right two places. The answer is seen to be −3.

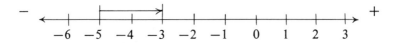

Do these with the number line.

a. $(-2) + 3 = ?$ c. $(-1) + 4 = ?$

b. $(-6) + 4 = ?$ d. $1 + 2 = ?$

The answers are a, $+1$; b, -2; c, $+3$; and d, $+3$. Be sure you can use the number line to add a positive number to either a positive or negative number.

Now, let's look at some patterns. You may wish to check these on the number line.

$4 + 2 = 6$	$2 + 2 = 4$	$1 + 5 = 6$
$2 + 2 = 4$	$1 + 2 = 3$	$0 + 5 = 5$
$0 + 2 = 2$	$0 + 2 = 2$	$-1 + 5 = 4$
$-2 + 2 = 0$	$-1 + 2 = 1$	$-2 + 5 = 3$
$-4 + 2 = -2$	$-2 + 2 = 0$	$-3 + 5 = 2$
$-6 + 2 = -4$	$-3 + 2 = -1$	$-4 + 5 = 1$
$-8 + 2 = -6$	$-4 + 2 = -2$	$-5 + 5 = 0$
$-10 + 2 = -8$	$-5 + 2 = -3$	$-6 + 5 = -1$

Study the patterns involving one positive addend and one negative addend. In every case, the answer can be obtained by *taking the difference between the addends and adding the sign of the addend which is most different from zero* (i.e., which is largest if you ignore the sign). Thus, $-1 + 5$ is $5 - 1$ with a plus sign, or $+4$; and $-6 + 5$ is $6 - 5$ with a minus sign, or -1. This "rule" is the standard way of doing additions like these.

Subtraction of a Positive Number from a Real Number

Subtraction can be defined as "moving left on the number line." Thus, "$6 - 4 = ?$" is defined as follows. Locate "6" on the number line and move *left* 4 units. Find the number corresponding to the stopping place. Do the example. The answer is, of course, 2.

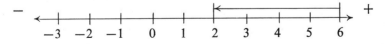

Let's do some more. What is $4 - 7$? This problem is undefined unless you have a number system with negative numbers. Locate "4" on the line and move left seven units.

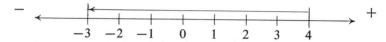

What is $(-1) - 1$? Start at "-1" and move left one unit.

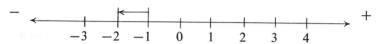

The answers are -3 and -2, respectively.

Now, let's look at some patterns. Use the number line to check these.

$$8 - 6 = 2 \qquad\qquad 6 - 4 = 2$$
$$6 - 6 = 0 \qquad\qquad 5 - 4 = 1$$
$$4 - 6 = -2 \qquad\qquad 4 - 4 = 0$$
$$2 - 6 = -4 \qquad\qquad 3 - 4 = -1$$
$$0 - 6 = -6 \qquad\qquad 2 - 4 = -2$$
$$-2 - 6 = -8 \qquad\qquad 0 - 4 = -4$$
$$-4 - 6 = -10 \qquad\qquad -2 - 4 = -6$$

Most of these subtractions follow the rule given in the previous subsection. However, a new case arises—the subtraction of a positive number from a negative number. The rule here is to add the numbers together and give the sum a negative sign ($-4 - 2 = ?$; $4 + 2 = 6$, so $-4 - 2 = -6$). In terms of moving on the number line, you start to the left of zero and move further left. The rule is easily clarified by thinking of the number line.

Addition of Negative Numbers to Real Numbers

The addition of a negative number to a real number can be handled by changing the sign of the number to a plus sign, and subtracting instead of adding. For example, $5 + (-2)$ is the same as $5 - (+2)$, or $5 - 2$. Also, $3 + (-5)$ is the same as $3 - (+5)$, or $3 - 5$.

This trick illustrates a basic principle: subtraction is merely the addition of a negative number. When you work in the real number system, which has negative numbers, you see that there really is no need to have an operation called "subtraction" because it is merely the addition of a negative number.

Subtraction of Negative Numbers from Real Numbers

The subtraction of negative numbers is a little more complex. It will be more readily understood after we take a look at multiplication of negative numbers. For now, let's be dogmatic and do the subtraction by rule only. The rule is to change the number to a positive number and add instead of subtracting. That is, change *both* the operation *and* the sign. Thus,

$$5 - (-1) = 5 + (+1) = 5 + 1 = 6$$
$$2 - (-4) = 2 + (+4) = 2 + 4 = 6$$
$$-8 - (-2) = -8 + (+2) = -8 + 2 = -6$$

Here are some patterns that will show that the above rule does make sense:

$$8 - (+4) = 8 - 4 = 4$$
$$8 - (+3) = 8 - 3 = 5$$
$$8 - (+2) = 8 - 2 = 6$$
$$8 - (+1) = 8 - 1 = 7$$
$$8 - (0) = 8 - 0 = 8$$
$$8 - (-1) = 8 + 1 = 9$$
$$8 - (-2) = 8 + 2 = 10$$
$$8 - (-3) = 8 + 3 = 11$$

In each of these examples, the minuend is a constant "8." The subtrahend decreases by one as we come down the rows. Therefore, the answers must always increase by one as we come down the rows, since we subtract one less each time. We see that the rule is necessary to get answers that keep the pattern consistent.

It might help to think of walking down the number line in the positive direction and turning around whenever you get a minus sign. In this analogy, the double minus sign would mean "turn around twice" and you are again facing in the positive (add) direction.

PROBLEM 6.2

a. $8 + 2$

b. $8 - 2$

c. $-8 + 2$

d. $-8 + (-2)$

e. $8 - (+2)$

f. $8 - (-2)$

g. $-8 - (+2)$

h. $-8 - (-2)$

i. $6 - 4 + 3 - 2$ m. $(1 - 3) + (2 - 3) + (4 - 3) + (5 - 3)$

j. $(9 - 1) - (-3 + 4)$ n. $(-7) - (-3) + (-2) + (-6)$

k. $(6 - 2) + (3 - 8)$ o. $-7 + 3 - 2 - 6$

l. $(-4 - 3) - (-6 - 1)$ p. $4 - 2 + (-6) - (-2)$

Multiplication of Real Numbers

TWO FACTOR MULTIPLICATION

There are three possible cases of multiplication with real numbers. These are a positive number times a positive number (which, of course, is what you already know quite well), a positive number times a negative number (or vice versa, which is still the same case), and, finally, a negative number times a negative number.

It helps to reconsider the nature of multiplication in general. Multiplication is merely repeated addition. So 6×3 is the same as $6 + 6 + 6$ or, equivalently, $3 + 3 + 3 + 3 + 3 + 3$. Multiplication involving a single negative number can be handled by this fundamental principle. For example, $3 \times (-6)$ is $(-6) + (-6) + (-6)$, or -18. Also, -6×3 is $(-6) + (-6) + (-6)$, or -18. Here are some more equalities: $3 \times (-6) = -6 \times 3 = -3 \times 6 = 6 \times (-3)$.

The minus sign can be considered as a third factor of -1, that is, $3 \times (-6) = 3 \times 6 \times (-1)$. All three factors can be commuted in several ways, each way being another way to write -18, for example, $3 \times 6 \times (-1)$, $(-1) \times 3 \times 6$, and $6 \times (-1) \times 3$.

A third way to look at the product of a negative and positive number is by patterns. Consider this set of problems:

$$6 \times 3 = 18$$

$$6 \times 2 = 12$$

$$6 \times 1 = 6$$

$$6 \times 0 = 0$$

$$6 \times (-1) = ?$$

In each row, the second factor decreases by one, so the product *must* decrease by six as we come down the rows. Therefore, the last entry must be six less than zero, or -6. We can thus continue as follows:

$$6 \times 0 = 0$$
$$6 \times (-1) = -6$$
$$6 \times (-2) = -12$$
$$6 \times (-3) = -18$$

The multiplication of two negative numbers is not as easy to demonstrate from the general definition of negative numbers. Let's do this one by patterns:

$$-6 \times 3 = -18$$
$$-6 \times 2 = -12$$
$$-6 \times 1 = -6$$
$$-6 \times 0 = 0$$
$$-6 \times -1 = ?$$

As we decrease the second factor by one in each row, the answer must *increase* by six. So the last entry must be six more than zero, or $+6$. We can continue the pattern:

$$-6 \times 1 = -6$$
$$-6 \times 0 = 0$$
$$-6 \times (-1) = 6$$
$$-6 \times (-2) = 12$$
$$-6 \times (-3) = 18$$

A very important special case is the product $(-1)(-1)$. Let's find the product, using patterns:

$$(-1)(3) = -3$$
$$(-1)(2) = -2$$
$$(-1)(1) = -1$$
$$(-1)(0) = 0$$
$$(-1)(-1) = ?$$

The answer must be zero plus one, or $+1$.

We can use our knowledge that $(-1)(-1) = 1$ and our ability to commute multiplication factors to get some additional insight into multiplying two negative numbers. As an example, consider $(-6) \times (-3)$. This can be rewritten as four factors: $(-1)(6)(-1)(3)$. Let's commute 6 and -1 to give $(-1)(-1)\,(6)(3)$. Since $(-1)(-1)$ is one, we have $(-6)(-3) = 6 \cdot 3 = 18$.

PROBLEM 6.3

See if you know how to multiply negative numbers now by doing these examples.

a. 5×5 e. $4(-3)$

b. $5 \times (-5)$ f. $-4 \cdot 8$

c. $(-5) \times 5$ g. $-4(-4)$

d. $(-5) \times (-5)$ h. $(-3)(+2)$

ANOTHER LOOK AT SUBTRACTION

Now we can return to the problem of subtracting negative numbers. The problem $6 - (-2)$ can be treated as $6 + (-1)(-2)$, or $6 + 2 = 8$.

MULTIPLICATION OF SEVERAL FACTORS

The preceding sections dealt with multiplying only two factors. But, several factors can be multiplied together, where some are positive and some are negative. For example, what is $(-6)(3)(4)(-2)(-3)$? Let's convert each negative number into a positive number times -1 and commute the entire system. We can write:

$$(-1)(6)(3)(4)(-1)(2)(-1)(3) =$$
$$(-1)(-1)(-1)6 \cdot 3 \cdot 4 \cdot 2 \cdot 3 =$$
$$(-1)(-1)(-1)(432) =$$

But $(-1)(-1) = +1$, so $(-1)(-1)(-1) = (+1)(-1)$, or -1, so the answer is -432.

Let's not spend a great deal of time here. It is sufficient to say that when the number of minus signs in a product of several factors is odd, the answer is negative. If there is an even number of minus signs, the answer is positive. For example:

$$(-2)(-2)(-2)2 = -16$$
$$(-2)(2)(2)(2) = -16$$
$$(-2)(2)(-2)(2) = 16$$
$$(-2)(-2)(-2)(-2) = 16$$

Division of Real Numbers

Since division is merely a form of multiplication, there is nothing particularly new here. Here is an example of each possible case.

1. $6 \div 2 = 6 \times (\frac{1}{2}) = 3$
2. $(-6) \div 2 = (-6) \times (\frac{1}{2}) = -3$
3. $6 \div (-2) = 6 \times (-\frac{1}{2}) = -3$
4. $(-6) \div (-2) = (-6) \times (-\frac{1}{2}) = 3$

In ratio notation, the same examples are as follows.

1. $6/2 = 3$
2. $-6/2 = -3$

3. $6/-2 = -3$
4. $-6/-2 = 3$

PROBLEM 6.4

Simplify by canceling.

a. $(-18)(2)/(-6)$
b. $[(-18)(2) + (14)(4)] \div [-4]$
c. $8(-3)(\frac{1}{2})(-6)(-\frac{1}{18})$
d. $[(-7)(-3)] \div [(-6)(14)]$
e. $-6 \cdot 7 + 5 \cdot 4 - 6 \cdot 3$

Exponents and Radicals Involving Negative Numbers

EXPONENTIATION

There is no particular problem involving the exponentiation of a negative number, since this is merely two, or several, factor multiplication. The sign of the product depends on whether the exponent is even or odd. Here are some examples.

1. $3^2 = 9$ $(-3)^2 = 9$
2. $3^3 = 27$ $(-3)^3 = -27$
3. $3^4 = 81$ $(-3)^4 = 81$
4. $3^5 = 243$ $(-3)^5 = -243$
5. $(-2)^2 + (-2)^3 = 4 - 8 = -4$
6. $3^2 + (-3)^3 + 3^4 = 9 - 27 + 81 = 63$

NEGATIVE EXPONENTS

Negative exponents are special division notation and were discussed in the previous chapter. We can show some patterns that illustrate the definitions of special exponents:

$$2^3 = 8 \qquad\qquad 3^3 = 27$$
$$2^2 = 4 \qquad\qquad 3^2 = 9$$
$$2^1 = 2 \qquad\qquad 3^1 = 3$$
$$2^0 = 1 \qquad\qquad 3^0 = 1$$
$$2^{-1} = \tfrac{1}{2} \qquad\qquad 3^{-1} = \tfrac{1}{3}$$
$$2^{-2} = \tfrac{1}{4} \qquad\qquad 3^{-2} = \tfrac{1}{9}$$
$$2^{-3} = \tfrac{1}{8} \qquad\qquad 3^{-3} = \tfrac{1}{27}$$

The rules for using negative numbers are also useful in what we have called exponential algebra. Here are some examples.

1. $(2^{-2})(2^{-3}) = 2^{(-2)+(-3)} = 2^{-5}$
2. $(3^2)^{-1} = 3^{(2)(-1)} = 3^{-2}$
3. $(2^4)^{-2} = 2^{(4)(-2)} = 2^{-8}$
4. $(2^{-3})^{-2} = 2^{(-3)(-2)} = 2^6$
5. $(2^3)(2^{-2})(2^{-1}) = 2^{3+(-2)+(-1)} = 2^0 = 1$

NEGATIVES UNDER RADICALS

Suppose you are working a numerical problem and it reduces to $\sqrt{-4}$. Where do you go from here? This is *not* the same thing as $-\sqrt{4}$, which is, of course, -2. The expression "$\sqrt{-4} = ?$" means, "Find a number such that when you multiply it times itself, you will get -4." Neither $+2$ nor -2 is such a number, because $(+2)(+2) = +4$ and $(-2)(-2) = +4$. The number $\sqrt{-4}$ cannot be handled in any way that has been previously discussed, since it is *not* an element in the real number system. Just as "$1 - 4 = ?$" cannot be answered without defining negative numbers, "$\sqrt{-4} = ?$" cannot be answered without defining numbers with the factor $\sqrt{-1}$.

Mathematicians handle such numbers by extending the number system to include them. This "augmented" number system is called the "complex number system." Numbers involving the square root of -1 are called "complex numbers." Sometimes they are called "imaginary numbers," but they are no less "real" than the elements of the real number system, so this is something of a misnomer.

Again, complex numbers occur relatively infrequently in statistical applications. Complex numbers arise in statistical theory problems and in some advanced applied techniques. However, complex numbers are rare even in the advanced, applied techniques.

Since complex numbers rarely are found in statistical applications, *usually* the finding of a value like $\sqrt{-4}$ means that a computational error has been made. For example, if you obtained $\sqrt{10(32) - 20^2} = \sqrt{320 - 400} = \sqrt{-80}$, chances are very good that there is an error in the calculations that gave you one or more of the numbers 10, 32, or 20.

Problems for Review

1. a. $36 + (-6)$
   b. $-42 + 7$
   c. $13 - 6$
   d. $-8 + 1$

   e. $(-3) + (-4)$
   f. $(-5) + (-16) + (-2)$
   g. $-5 + 3 - 2 - 4 + 1$
   h. $(-6) + (-1) + (-3) + (-2)$

2. a. $-6 - 2$
   b. $-6 - (+2)$
   c. $-6 - (-2)$
   d. $17 - (-4) - 2$

   e. $72 - 43 - 13$
   f. $(-6) - (-2) - (+4)$
   g. $-(-4) - (-1)$
   h. $7 + (-1) - (-4) + 2$

3. a. $16 \cdot 2$
   b. $-4 \cdot 3 \cdot 2$
   c. $(-4)(-3)(-2)$
   d. $(-7)(4)(-2)(3)$

   e. $(-4)(-5)(+4)$
   f. $(-4)^2 + 4^2$
   g. $(-5)^3(-2)$
   h. $(-6)(-2)(-3)^2$

4. a. $(-6)(-3) \div 4 \cdot 3$
   b. $(-72) \div (-8)(-3)$

   c. $(-6)^{-1}(-3)^{-1}(-1)^3$
   d. $-(4^2) \div (-4)^2$

5. a. $(-6)(3) + 6 \cdot 4 + (-2)(-3)$
   b. $(-6)(4) - (-1)(-3)$
   c. $10(36) - 15^2$

d. $(5 - 3) + (2 - 3) + (4 - 3) + (1 - 3)$

e. $(5 - 3)^2 + (2 - 3)^2 + (4 - 3)^2 + (1 - 3)^2$

f. $(0 - 2)(6 - 4) + (1 - 2)(5 - 4) + (3 - 2)(3 - 4)$
$+ (4 - 2)(2 - 4)$

g. $6^2 + 5^2 + 3^2 + 2^2 - (6 + 5 + 3 + 2)^2/4$

h. $\sqrt{4(124) + (-12)(20)}$

Part Two

BASIC ALGEBRA REFRESHER

THIS part is a review of algebra topics as they relate to the study of statistics. The emphasis here is on the fundamental ideas and the language of algebra and on the algebraic manipulation of symbols.

The knowledge of terminology is necessary in the understanding of written and oral communication of mathematical formulas. The symbols and algebraic operations are not only invaluable mathematical tools but also are extremely important in reducing the memory load in statistics learning.

The statistics student is often given several formulas for a single concept. Often these formulas are merely different ways of writing the same expression. If he does not know how to handle the algebra required to reveal the relationships between the formulas, he is at a disadvantage. If he must memorize several versions of a single formula because he cannot do algebra, he loses valuable time that could be spent studying basic statistical principles. Moreover, if he can handle the algebra then, he needs to memorize only one form of the formula and, from the one form, he can "derive" other versions.

For example, the formula for relating the circumference (C) of a circle to the radius (r) of a circle can be written in two ways—$C = 2\pi r$, or $r = C/2\pi$. With a little knowledge of algebra, we can treat both expressions as equivalent —only one needs to be memorized.

Statistical formulas can often be written in relatively simple ways *and* in relatively complex ways. Students who know algebra can memorize the simple formulas and use algebra to determine the more complex formulas. Whereas, students who know little algebra often resort to attempting to memorize both simple and complex versions. This extra work can prevent them from having sufficient time to study fundamental principles and applications—they are too bogged down in the formulas. Thus, they lose in several ways.

This section, then, has several goals. The more important goals are to learn standard terminology, some new "operations" that are necessary in statistical calculations, and common algebraic operations.

Variables, Constants, and Parameters

Introduction

THE purpose of this chapter is to introduce some algebraic language. The topics are fundamental, yet they can be a source of unusual perplexity and confusion when not understood. The use of symbols other than arabic numerals will be introduced. Also, words and terms will be explained.

There will be no calculations to learn; however, don't let this fact fool you into glossing over this material. Mathematics is supposed to be a science of logical thinking, and logical thinking is difficult if terminology and fundamental principles are mastered poorly.

Things and Names of Things

Let's start our algebraic studies with a very fundamental proposition: an object and the name of the object are two different things. Now, no one would argue that this book and the letters "b," "o," "o," and "k," taken in the given order, are the same things. One is a concrete object. The letter combination is a name, a set of symbols, a label. The object is called the *referent* of the symbol "book." Certainly there is no problem distinguishing the symbol "cow" from its referent.

However, in mathematical writings and understanding, often there is considerable confusion between symbols and their referents. Let's try to keep clear the distinction between a symbol and its referent.

Throughout this book, we will use a symbolization that is almost standard ("almost" in the sense that it usually is used by persons who are careful in distinguishing symbols from referents—many writers are not). When we wish to discuss a symbol as a symbol, it will be enclosed in quotation marks. When there are no quotation marks, then what is meant is the referent of the symbol. Here are some correct examples.

1. A cow is a mammal. How do you pronounce "cow?"
2. Soft-covered books are generally called "paperback books."
3. The word "friend" is misspelled in your paper. John is my best friend.
4. There is too much confusion among the use of "to," "too," and "two."

Hopefully, this text will be consistent in the use of quotation marks to designate names.

One source of difficulty in algebra often is the student's confusion in the distinction between numbers and the symbols for numbers. Symbols for numbers are called "numerals." When we wish to talk about a number, we will write a numeral. When we wish to talk about a numeral, we will write the numeral and put quotation marks around it. Here are some correct examples.

1. Two ways to write "two" are "2" and "1 + 1."
2. "1,000,000" has six "0's." 1,000,000 is 10^6.
3. The statement "2(1 + 3) = 5" is wrong; it should be "2(1 + 3) = 8" because 1 + 3 = 4 and 2 × 4 = 8.

The number-numeral confusion arises largely because numbers are abstract concepts. You cannot point to a number as you can a book or a cow. The referent of the numeral "5" is a *concept* that we might call "fiveness," which is developed over years of experience with numbers. Fiveness becomes understood by a child largely with the aid of finger and toe counting.

Why should we bother with this discussion? It is because algebra introduces some unusual symbols and notations that are clarified by distinguishing symbols from their referents. Consider the number five. It has many symbols associated with it, all of which are equivalent in the sense that *each has the same referent*. Here are several of the infinite number of ways to write a symbol for the number five.

Five	$4 + 1$
V	$\sqrt{25}$
卌	$55 \div 11$
$710 - 705$	The remainder of $65 \div 12$

Each of these expressions has the same referent. They are all different names for the same thing.

PROBLEMS

Punctuate the following sentences by inserting quotation marks where appropriate.

7.1. Why did you name a white dog Blacky?
7.2. What is the sum of two plus two?
7.3. What is the purpose of the 2 in the expression $3^2 = 9$?
7.4. There is more than one way to write two. There are an infinity of ways to write two.
7.5. Counting numbers include 1, 2, 3, 4, 5, and 6.
7.6. The vowels are a, e, i, o, u, and sometimes y.
7.7. The answer is not 7.
7.8. How do you write one billion?
7.9. Can you count by fives?
7.10. VI is the Roman numeral for six.
7.11. VI and 6 mean the same thing.

Letters as Names of Numbers

Since numerals are human inventions, the use of a particular numeral for referring to a particular number is somewhat arbitrary. For example, I could use the letter "*a*" as a symbol for one and the letter "*b*" as a symbol for two. Then I could say things like

$$\text{"}a + a = b,\text{"} \quad \text{"}b - a = a,\text{"} \qquad \text{or} \qquad \text{"}(b)(a) = b\text{"}$$

Each of these sentences is true and is meaningful *in context*. Arabic numerals are necessarily used in order to be able to communicate, but there should be nothing mysterious about using nonnumeric symbols for numbers.

The common equation for relating the circumference of a circle to the radius of the circle is $C = 2\pi r$. This formula includes the Arabic "2," the Greek letter "π," and the letters "C" and "r." The "2" refers, of course, to the number two. The "π" refers to a number that is close to, but not exactly equal to, $\frac{22}{7}$. It is not a rational number, that is, it cannot be written *exactly* in decimal *or* ratio notation; therefore, notation other than decimal notation is used—the Greek letter pi. The number pi is no less a number than two or four. It is merely inconvenient that it cannot be written exactly in decimal or ratio notation, so a letter is used for convenience to denote pi.

The letters "C" and "r" are also used for convenience. By using them, the formula becomes applicable for any circle, which is certainly more convenient than needing an infinite number of equations, one for each possible circle size. The letters "r" and "C" refer to specific numbers wherever the formula is used in the context of an application.

Basic Definitions

INTRODUCTION

This section defines many terms used extensively in mathematics and statistics. An important consideration in the use of any of these terms, just as in the use of any word, is the *context* in which the word or term is used. Keep the context in mind, and you can avoid some of the confusion that many students have. In particular, whether or not a numeral or other symbol refers to a constant or not, or whether it refers to a parameter or not, is a matter of context. A "5" can refer to a specific value of a variable, a known constant, a parameter that is known, or any of several things, depending on the context in which it is used.

The first several subsections deal with distinctions between constants and variables. It is unfortunate that the use of the terms "constant" and "variable" is so standard, since they connote movement versus nonmovement or change versus nonchange, but only in relatively rare instances is there actual movement or change related to a mathematical variable in the behavioral sciences. The terms "variable" and "constant" are borrowed largely from physics where it makes sense to talk, for example, about an object falling in space with time (a variable) constantly increasing, distance fallen (a variable) constantly increasing, and the force of gravity acting on the object with a constant effect.

CONSTANTS

Definition. The word "constant" refers to symbols or numerals that have only one number for their referent. Again, you must keep context in mind when you used the word "constant."

Arabic Numerals Used to Denote Constants. Usually, constants are symbolized by Arabic numerals. Some texts even erroneously define "constants" in this way. In the formula $C = 2\pi r$, the symbols "2" and "π" refer to numbers that are constants in the formula. Here is another expression: $y^2 + 3y + 6$. The constants in this second example consist of 2, 3, and 6, corresponding to Arabic numerals "2," "3," and "6."

Letters Used to Denote Constants. Letters can also be used as constants. For example, we can multiply 2 times π and use "k" as the symbol for 2π. Then "$C = 2\pi r$" can be written "$C = kr$," in which the "k" refers to the constant 2π. The symbol "π," of course, is also a letter, but it is a Greek letter.

Here is another example. Consider this set of expressions:

$$0 + x = 1$$
$$1 + x = 2$$
$$2 + x = 3$$
$$3 + x = 4$$
$$4 + x = 5$$

If all of these statements are *simultaneously* true, then "x" must refer to a *constant*—namely, the number one.

So, a constant can be either an Arabic numeral, a letter, or any other symbol as long as, in context, the *symbol has a single number as its referent*.

VARIABLES AND RANGES OF VARIABLES

Definition. The word "variable" refers to a symbol with more than one number as a referent. The term "range of a variable" means the set of all referents to which the variable might refer. Usually, the context will determine the range of a variable. Here is an example: $y = x^2$. If nothing else is known but that y and x are variables referring to real numbers, then one can infer that x has as its range *all real numbers*, while y has as its range *all positive real numbers* plus zero. (Since a negative number squared is always a positive number, y cannot be negative.) However, the context of the problem might specify that numbers to which x can refer consist only of integers between 0 and 5; then the variable x has a range of $\{0, 1, 2, 3, 4,$ and $5\}$, whereas the variable y has a range of $\{0, 1, 4, 9, 16,$ and $25\}$.

Letters Used to Denote Variables. Usually letters are used as symbols for variables. However, it is important to remember that letters can also refer to constants. But this presents no particular problem if you consider that a constant is merely a special case of a variable—a constant is a variable with a range consisting of only one number.

PARAMETER

The term "parameter" is difficult because it refers to numbers that are simultaneously constants from one point of view and variables from another.

Let's consider an example. Let's define a variable and call it "x." The variable x is a score on an inventory of mathematics background. Scores are defined as follows:

$x = 0$ if the interviewee had no high school math
$x = 1$ if the interviewee had only one math course
$x = 2$ if the interviewee had two courses
$x = 3$ if the interviewee had more than two courses

So the range of x is 0, 1, 2, and 3. Now let's define four groups of college freshmen, using the variable x. Let's assume that each of these four groups will be assigned to different college math sequences. Across all groups, x is a variable, but *within the groups*, x is constant. We can call x a parameter— it is a number *unique to the subgroup*, but which has different values over *all* students. In this case, the *parameter* actually *defines the subgroups*.

Here is another example. The equation $X + Y = A$ is a general expression for all pairs of numbers X and Y that add up to A. A, X, and Y are all variables; however, A is also a parameter. If a particular number is substituted for A, one particular set of X and Y pairs is possible. A different number for A would lead to different X and Y pairs. Thus, A is a parameter since it defines the (X, Y) sets. Here are some examples of pairs of X and Y for three values of the parameter A.

$A = 0$		$A = 2$		$A = 4$	
X	Y	X	Y	X	Y
−1	1	1	1	2.5	1.5
−2	2	2	0	4	0
−3	3	0	2	−9	13
2	−2	−4	6	7	−3
10	−10	3	−1	2	2

You can see how a parameter acts as both variable and constant. A is a variable in the sense that any number is in its range. However, for a particular set of the (X, Y) pairs, A is constant. It is a *parameter* that defines the elements in the sets of (X, Y) pairs.

COEFFICIENTS

The word "*coefficient*" is used often in discussing the symbols in an algebraic expression. It can be considered as a synonym for the word "factor." Consider the equation

$$y = 5x + 1$$

The symbols "$5x$" can be used as an example. We can say "5 is the coefficient of x," or "x is the coefficient of 5," since both 5 and x are factors of $5x$.

Traditionally, "coefficient" was used only for nonvariables, and some texts might still use this old definition. That is, in "$y = 5x + 1$," some texts might say that only "5" is a coefficient since "x" refers to a variable. However, the modern usage of "coefficient" allows us to call "5" a coefficient of x and call "x" a coefficient of 5.

Consider this algebraic expression:

$$5ax$$

The coefficient of x is $5a$, the coefficient of a is $5x$, and the coefficient of 5 is ax.

PROBLEM 7.12

In the equation $y = 42b + cX - 25aZ$, what is the coefficient of:

a. b e. Z
b. c f. aZ
c. X g. 42
d. a h. y

EQUATIONS AND INEQUALITIES

Equations and inequalities are mathematical sentences. They are statements like any good English declarative sentences; and like any declarative sentences, they can be true or they can be false. Here is an example of an equation that is a true sentence:

$$[7 + 2(10)]/3 = 9$$

Here is an example of an equation that is not true: $2 + 2 = 5$. Both are perfectly good statements, but one is true and one is not.

An *equation* is a mathematical sentence that specifies that two numbers are equal. It always contains an equal sign ($=$). Or, in other words, an equation specifies that two expressions are names for the same number. The equation

$$[7 + 2(10)]/3 = 9$$

says that "$[7 + 2(10)]/3$" and "9" have as their referents the same number (namely, 9).

The equation $X^2 + B = C + 1$ means that whatever the referent of X, B, and C, "$X^2 + B$" and "$C + 1$" are equivalent names for the same referent number. Again, the equation may be true or false, just like any English sentence.

An inequality also is a sentence that is either true or false. There are five inequality signs to know. These are "$\neq$," ">," "<," "$\geq$," and "$\leq$."

The symbol "$\neq$" means "is not equal to" and is illustrated by the true sentence "$2 + 2 \neq 5$." The symbol ">" is read "is greater than." The true sentence "$5 > 2 + 2$" is read, "Five is greater than $2 + 2$." The symbol "<" is read "is less than." The true sentence, "$2 + 2 < 5$" is "two plus two is less than five." Notice that the arrow in the inequality sign always points to the smaller quantity. Here are some more examples of true inequalities. Assume a and b are positive numbers:

$$7 + 2 > 8 \qquad 1 + 1 \neq 3$$
$$\sqrt{3} > 1 \qquad \sqrt{a + 1} > \sqrt{a}$$
$$\sqrt{3} < 2 \qquad ab < a(b + 1)$$

The two examples "$\sqrt{3} > 1$" and "$\sqrt{3} < 1$" can be combined into a compound sentence "$1 < \sqrt{3} < 2$" which can be read "$\sqrt{3}$ is less than 2 *and* it is greater than 1" or "$\sqrt{3}$ is *between* 1 and 2." The latter reading is preferred since it is clearer and briefer. Also notice that the last two sentences are true without regard to the numbers to which "a" and "b" refer (as long as they are positive numbers).

The symbols "$\geq$" and "$\leq$" are read "is greater than *or* equal to" and "is less than *or* equal to," respectively. Here are some examples:

$$7 + 2 \geq 8 \qquad \sqrt{24} \leq 5$$
$$7 + 2 \geq 9 \qquad \sqrt{25} \geq 5$$
$$9 \leq 7 + 2 \qquad \sqrt{25} \leq 5$$
$$9 \leq 17 + 1 \qquad a + 10 \geq 10 \text{ if } a \geq 0$$

Equations and inequalities are of utmost importance. For this reason an entire chapter (Chapter 9) is devoted to these topics.

MODELS

The word "model" is a generic term used in research to denote any of a variety of explanations of phenomena. Many models are expressed mathematically as equations or functions (discussed next), so it is appropriate to discuss models in this chapter. Some models are merely verbal descriptions of phenomena; some are sophisticated mathematical theories.

At one extreme we might say, "Behavior is a product of heredity and environment," while at the opposite extreme we might say, "The total energy

(E) contained in any physical object is expressed by $E = mc^2$, where m is a parameter depending on the mass of the object and c is a constant." Usually statisticians try to formulate explanations of research results or expected research results in some sort of equation involving variables, constants, and parameters.

The model that is the most commonly used statistical model is the normal (or Gaussian) equation. When this is graphed, it has a bell-shaped appearance. The model is used often in grading ("grading on the curve"), in applying test scores, and in statistical inference. The equation can be written

$$P = (1/\sqrt{2\pi})e^{-(x-\mu)^2/2}$$

where P and x are variables, μ is a parameter, and all other symbols refer to known constants. More general, or complex, versions can be written that involve other subgroup parameters. Both "π" and "e" refer to known constants that are irrational numbers (i.e., they can't be expressed by an Arabic numeral in decimal or ratio notation). The equation is a model for the frequency with which x takes on the various values in its range. The parameter "μ" corresponds to the subgroup average score. The problem of the statistician is to estimate this parameter for various subgroups.

Another common model is the linear model. This is an equation that says an event (variable) is determined largely by a weighted sum of other variables. The linear model is used extensively in prediction problems. The "weights" are parameters associated with the particular group of persons or things for which predictions will be made. The statistical problem in prediction is to estimate these parameters.

FUNCTIONS, CORRESPONDENCE, AND ORDERED NUMBER PAIRS

Consider this set of pairs of numbers. If you continued the list in the same pattern, what two numbers should come next?

$$(1, 1)$$
$$(2, 4)$$
$$(3, 9)$$
$$(4, 16)$$
$$(5, 25)$$
$$(?, ?)$$

The pattern clearly indicates that the first numeral should be "6" since the first column is "1, 2, 3, 4, 5, ?" The second column consists of numerals each of which *corresponds* to the square of the number indicated by the first column numeral. The numeral pairs are *ordered*, that is, the second numeral refers to the square of the number referred to by the first numeral. We can say that there is a *correspondence* between column one and column two and this correspondence is defined by the rule, "The second number is the square of the first number." Here it is important to keep the numbers in order. The number pair (36, 6) does not belong in the list anywhere because it is not in the order specified by the rule.

Rules like the one above have a special name. They are called "*functions.*" Functions are rules that establish a correspondence between number sets. In an ordered pair, the function is the rule by which one number can be determined by the other.

Another rule or function is, "The second number in the pair is 4 times the first number minus 3." Here are some of the number pairs based on this function!

$(-2, -11)$	$(2, 5)$
$(-1, -7)$	$(3, 9)$
$(0, -3)$	$(4, 13)$
$(1, 1)$	$(5, 17)$

Both of the functions used as examples can be expressed by equations relating two variables. These equations are merely shorthand ways of writing the function or rule of correspondence. For the first function, let's use "y" to symbolize the first number in each pair and let's use "x" to symbolize the second number in each pair. Then each pair (x, y) is defined by the equation, or mathematical function, $y = x^2$.

We can say that the second example consists of all (x, y) such that $y = 4x - 3$. The equation "$y = 4x - 3$" is the mathematical function by which one can determine what number y corresponds to any x.

Each of the previous examples has one number corresponding to exactly one other number, so we call these examples of *one-to-one correspondence*. We can also have many-to-one correspondence, one-to-many correspondence, and even many-to-many correspondence. The square root function establishes a one-to-two correspondence, e.g., $\sqrt{4} = +2$ *and* -2; $\sqrt{25} = +5$ *and* -5; etc.

Psychometric research almost always involves one-to-many correspondences. Each subject (person) studied has many psychometric variables associated with him. The functions in this case are quite complex and frequently are not defined mathematically but in a careful description of the data-gathering process.

Let's consider another example. Here are some unordered sets of number pairs: $(-2, 2)$, $(1, -1)$, $(3, -3)$, $(-7, 7)$, etc. Each number in each pair is the negative of the other, but there is no order in the pairs. Now, let's establish a correspondence between these sets of unordered numbers and another number by the function $b = \sqrt{a}$. Let "b" represent all numbers such that $b = \sqrt{a}$, and let "a" represent all *positive* real numbers. In this example, each number in the range of a gives rise to two numbers for b. We can write some examples of the *ordered* sets (b, a) as:

$$(1, (1, -1)) \qquad\qquad (9, (-3, 3))$$
$$(4, (-2, 2)) \qquad\qquad (16, (4, -4))$$

Here, we have a one-to-two correspondence. Each number, a, corresponds to two numbers, b, and the order of the two numbers, b, is of no interest.

Here are some functions that might be used in behavioral research.

1. Here are the names of several children. Beside each mark "1" if it is a boy and mark "0" if it is a girl: John, Mary, Joe, etc. (This is a many-to-one correspondence, that is, for the one value "0," there will be many children; for the one value "1," there will be many children. The pairs will be (John, 1), (Mary, 0), (Joe, 1), etc.)

2. Compute a prediction, y, by multiplying X by 1.32 and adding 5.4. (The function is $y = 1.32X + 5.4$; this is a one-to-one correspondence.)

3. If the child's test score is above 100, send him to classroom 10, if his score is between 50 and 100, send him to classroom 8, and if his score is 50 or less, send him to classroom 4. (This "function" assigns a room number to each test score. It is also a many-to-one correspondence, or a one-to-many correspondence, depending on how you order the sets.)

PROBLEMS

The mathematical (and statistical) problems to learn from this discussion are how to formulate verbal correspondence rules into mathematical functions and, conversely, how to interpret mathematical functions in terms of understandable English. Write an equation for each of these examples.

7.13. Let d be a number obtained by squaring the difference between a and 60.

7.14. Let k be a number such that it is smaller than j plus 10, but larger than $j - 10$.

7.15. From each number X subtract 50. Divide this difference by 10. Call the result Z.

7.16. Find the function that defines the number pairs (x, y) if these are examples of such pairs:

$$(1, 1.1) \qquad (6, 1.6) \qquad (10, 2) \qquad (100, 11)$$
$$(4, 1.4) \qquad (.5, 1.05) \qquad (20, 3) \qquad (1000, 101)$$

Function Symbolization. Sometimes it is helpful to use a special function notation. Consider the squaring function. Instead of "$y = x^2$," frequently you will see "$f(x) = x^2$." This is read "f of x is x-squared," or "the function of x is x-squared." The "$f(\ )$" symbol is shorthand for "function of." Any other letter can also be used to denote functions. The parentheses are the important key to the symbolization. We could have written "$g(x) = x^2$," or even use the Greek alphabet as "$\Psi(x) = x^2$." Context will generally determine whether or not parentheses indicate functional relationships or multiplication.

Functions of Functions. Sometimes it helps to chain several functions together. For example, let $y = x^2$. Also, let $Z = y + 5$ be a second function. This Z can be written directly as a function of x: "$Z = x^2 + 5$." However, sometimes it is convenient to keep the two operations as separate steps. In the parentheses notation we can say $y(x) = x^2$ and $Z(y) = y + 5$. Putting these together, Z is a *function of a function*, and we can write

$$Z(y) = Z(y(x)) = y(x) + 5 = x^2 + 5$$

Here is a statistical example. Let μ and σ be constants. Let X be a variable of interest. Define $x(X) = X - \mu$. Also define $Z(x) = x/\sigma$. Then

$$Z(x) = Z(x(X)) = x(X)/\sigma = (x - \mu)/\sigma$$

In the first example, the symbols "$Z(y(x))$" show that Z is a function of a function. In the second example, "$Z(x(X))$" says that Z is a function of a function.

Numbers Series and Special Notation

Introduction

THIS chapter introduces several related concepts. One of the concepts deals with the notation used to distinguish between numbers that are similar in some ways, but different in other ways. These notation systems include subscripts, superscripts, primes, and other special marks. This notation frequently presents statistics students with unnecessary difficulties which we hope to help him avoid.

A second set of problems introduced are special addition and multiplication operators that are used very commonly in statistical work. It will be extremely helpful to you to learn to use these operators now so that you will not need to study these skills during your statistics courses. The applications in this chapter are heavily oriented to actual statistical calculations.

Numbers in a Series

The previous chapter introduced the concept of "ordered pairs of numbers." This concept is easily extended into larger sets of numbers whose order is extremely important. Numbers can be ordered by numerous criteria. They can be ordered by magnitude from low to high, or high to low. They can be ordered by some complex mathematical function or by simple mathematical functions. Here are some sets of numbers ordered by simple functional relationships. The three dots mean that the set can be extended indefinitely.

1. 1, 2, 3, 4, 5, 6, . . .
2. 2, 4, 8, 16, 32, 64, . . .
3. 0, 5, 10, 15, 20, . . .

The functions are, respectively:

1. Each number is one plus the number preceding it.
2. Each number is two times the number preceding it.
3. Each number is five more than the number preceding it.

Here are some examples of fairly complex number series and the rules by which they are constructed. Let x be any number in the sequence. Then let n be the next number in the sequence. The number n can be determined by x and the rules that are given.

1. Rule: $n = (2x + 1)^2$:

   $0; 1; 9; 361; 522,729; \ldots$

2. Rule: $n = (x + 1)$ if x is even or zero

   $n = 2x$ if x is odd:

   $1; 2; 3; 6; 7; 14; 15; 30; 31; \ldots$

3. Rule: $n = x^2/2$:

   $1; 1/2; 1/8; 1/128; 1/32,768; \ldots$

You can check these series for accuracy by applying the rules.

Sometimes the numbers are in a regular pattern, as in the above samples. However, in statistical work we frequently deal with number series that have no particular pattern.

Let's call a sequence without a pattern a "*random number series.*" The concept of "randomness" receives a large emphasis in any statistics course where you will be given more accurate definitions. Let's just say that the term "random number series" will refer to a series of numbers, none of which is a mathematical function of any other numbers in the series.

Let's consider a concrete example. Suppose we conduct a poll and ask twenty persons on a busy street to guess the percentage of voters in the next state election who will vote for the Democratic candidate. If we do not ask for names and merely record the responses on a tablet, we will get a series of numbers that probably will be a random series, according to our definition.

Suppose we are recording IQ scores out of children's school records, which are stored alphabetically. Our series of IQ scores can also be considered as a random series. Although the list is certainly in a predetermined order, the numbers themselves will not have any particular pattern.

Subscript Notation

The last two examples of random number series were examples of standard statistical data. In the election poll example we were not interested in identifying the individuals responding, so the order was entirely irrelevant. However, in the IQ recording example, we might be able to use the IQ scores in guidance or teaching, so we might wish to record some identification along with the IQ scores. Let's be mathematical and use identification numbers instead of names. Instead of using school identification numbers or social security numbers, let's simply number the records from one up to N, where N is the total number of children whose IQ's were recorded. We are establishing a one-to-one correspondence between the IQ's and the identification numbers. Also, we need to name our IQ variable. Let's just call it "X."

If we wish to talk about a specific individual's IQ, we can do it with our symbol "X" and a subscript. (Subscripting is standard notation and *must* be learned.)

We can refer to the first child's IQ by the symbol "X_1," which is our label for IQ and identification number "1" (the subscript "1"). "X_1" means "the first X." Similarly, the second child's IQ is called "X_2," the third is "X_3," and so on. Let's look at a small example of such a list (Table 8.1).

Table 8.1

Identification Number	Name	IQ	Subscript Symbol
1	Adams	93	X_1
2	Baker	113	X_2
3	Conway	134	X_3
4	Davis	75	X_4
5	Edwards	90	X_5
6	French	102	X_6

A symbolic way of describing the data is, "Let X_i be the child's IQ, where i takes on the values of 1, 2, 3, 4, 5, and 6," and an even more symbolic way of saying the same thing is, "Let X_i be the child's IQ, where $i = 1, \ldots, 6$." The set of symbols "$i = 1, \ldots, 6$" means, literally, "i is equal to the numbers 1, 2, 3, $\ldots$ and so on up to 6, taken in that order."

The lowered numerals are *subscripts*. The subscript "i" denotes a general *element* in the list, i.e., any one of the IQ's or X's, no particular one. Any letter can be used as a general subscript; however, it is common to use the letters i, j, k, and l more often than other letters. The letter "N" is often used in statistics to identify the highest subscript. It is shorthand for "Number" and usually means, "The number of persons or things studied."

PROBLEMS

Here is a series of random numbers called "X." Reorder them from low to high and assign identification numbers to them where "1" indicates the smallest number in the series and "10" indicates the largest:

$$84, 17, 34, 12, 95, 62, 29, 30, 74, 14$$

8.1. Which number now corresponds to the following?

a. $X_3 = ?$ b. $X_6 = ?$ c. $X_9 = ?$

8.2. What is the sum of X_1, X_2, and X_3?

8.3. What is $X_9 - X_7$?

8.4. Suppose you had not reordered the number series. Suppose you assign identification numbers to the data in the order that they appear above ($X_1 = 84, X_2 = 17, \ldots, X_{10} = 14$). What "new" labels are associated with each of the previous problems? (That is, rewrite the problems in terms of the new subscripting so that the numerical answers are the *same* as before.)

Superscripts, Primes, and Other Special Notation

This section describes standard ways to construct symbols for variables (or constants). In particular, labeling techniques for distinguishing between numbers that have common properties, yet differ slightly, are described. The previous section on subscripting could be considered as one example.

MORE ON SUBSCRIPTS

In the previous sections, subscripts were used to distinguish between numbers in a series. You can think of the numbers in the series as a set of particular values to a variable. Thus, the numbers all have something in common, namely, they are all values of a particular variable. But they are each different in that each corresponds to a different identification number. Thus, we used subscripts to distinguish among a group of numbers that were alike in some ways, yet unlike in other ways. This is a common research technique.

Suppose, for example, we have tested one child three times with an arithmetic test. We could name his three arithmetic scores A_1, A_2, and A_3, where the subscripts mean "first testing," "second testing," and "third testing," respectively.

Another common psychometric example is "corrected-for-guessing scores." We can call the uncorrected score "X" and call the corrected score "X_c." The subscript c would be used as shorthand for "corrected." It is not a variable; it is not even a symbol for a number. It is a *letter* distinguishing between the corrected score (X_c) and the uncorrected score (X). The letter "X" is used in both cases to show that "X" and "X_c" both refer to the same variable. The subscript "c" shows that "X_c" and "X" are not the same number. They are alike in one very important way, and yet they are unlike in one very important way.

Problems

Assume that quantitative definitions exist for S, O, and E as defined below:

S = socioeconomic status index
O = occupational level
E = educational level
j = identification number subscript ($j = 1, \ldots, N$)
$N = 25$
a = adjusted for educational bias

Write good symbol combinations using subscripts for the following sentences.

8.5. The unadjusted socioeconomic status of the third subject is 9.

8.6. The unadjusted socioeconomic status equals the occupational level for all persons.

8.7. The adjusted socioeconomic status equals the occupational level plus two times the educational level for all persons.

8.8. The difference between adjusted and unadjusted socioeconomic status equals twice the educational level for all persons.

Superscripts

Just as "c" was used as a subscript for "corrected-for-guessing," it could be used as a superscript for the same purpose: "Let 'X_i^c' denote the corrected-for-guessing score of individual i." This nomenclature prevents the double subscript "X_{ci}." In this case, there is no problem in confusing "c" with an exponent since "c" does not refer to a number; it refers only to the word "corrected-for-guessing." However, numerals are also used for superscripts, but this is not too common. Subscripts are constantly used in applied statistics. Superscripts are used often but not as commonly as subscripts. They are used more commonly in theoretical work in which many symbols are needed.

PRIMES

Primes (') are also commonly used to distinguish several different numbers that are similar in some way. Thus, we can write three daily temperature measurements with subscripts as "T_1," "T_2," and "T_3." Or, we can use prime notation and say T', T'', and T'''. (In this sentence we are violating the quotation mark rule for symbols, but the use of quotation marks here would be quite confusing.)

Here is an example. Let Y be a variable we have measured. Let Y' be an estimate of the same variable based on a different measuring instrument. Y and Y' are similar, yet different, numbers.

TILDES

The *tilde* (~) also is used for the same purpose. We might have three measurements of Y. We can denote these by "Y," "Y'," and "$\tilde{Y}$."

CARATS

The *carat* (^) is used similarly. We could distinguish four measurements of Y as "Y," "Y'," "$\tilde{Y}$," and "$\hat{Y}$."

SIMILAR LETTERS

Another common way of writing related, yet different, labels for numbers is with similar letters. For example, the letter "s" can be used as a label for a variable, and if there is need for four similar "s's" to distinguish between related, but slightly different variables, we could use these symbols "s," "S," "Σ," and "σ." (Σ and σ are the capital and small case Greek s.)

Here is a common example. Let "X" be a name for a variable. Let "A" be the name for the *average* value of X. Then, let $x = X - A$. Statistical researchers often have need to distinguish between a value of a variable and the deviation of this value from the average. This distinction is usually made by using capital letters for the basic data and small-case letters for deviations of the basic data from the average.

COMBINATIONS

It should be clear that any and all of these symbols can be used together in combination. However, subscripts (and, to a lesser extent, superscripts) provide the usual technique for distinguishing between numbers in a series.

The use of primes, carats, and tildes are especially common in naming statistical formulas. If several formulas provide slightly different ways to calculate estimates of a quantity, then these formulas frequently are made distinct by using the special symbols.

For example, the concept of "standard deviation" is a first-course concept. Variations of the letter s are used to distinguish among different formulas for the standard deviation. Some of these are s, $\hat{s}$, Σ, and σ. We often see various sub- and superscripts attached. Thus, we might see the statement, "s_c is a better estimate of σ than s is." Each symbol in that sentence perhaps refers to "standard deviation," yet each refers to a slightly different concept or formula concerning the standard deviation.

Context and familiarity with an author's nomenclature will keep these straight. An author usually provides "ground rules" which help. Thus, he *might* say, "Greek letters will always indicate *theoretical* values which we wish to estimate; Roman letters will always indicate *estimates* of theoretical values. Small-case letters will always indicate deviations-from-averages; etc." You must pay strict attention to the author's symbolization if you are to understand the text.

The Summation Operator, Σ

THE BASIC DEFINITION OF "Σ"

Adding accounts for a great deal of computational time in statistics. The special operator "Σ" is used in conjunction with summation of subscripted (or superscripted) number series. The symbol "Σ" is read "the sum of . . ." or "add up . . . ." In particular, you will see the summation "Σ" usually used with subscripted variables in statistical equations.

Let's work with an example. Let's define a variable X_i ($i = 1, \ldots, 6$) as follows:

$$X_1 = 10 \qquad X_4 = 40$$
$$X_2 = 20 \qquad X_5 = 50$$
$$X_3 = 30 \qquad X_6 = 60$$

The symbol to designate the command, "Find the sum of all of the X_i starting with X_1 and going to X_6," is

$$\sum_{i=1}^{6} X_i$$

The symbol means, literally,

$$\text{"}X_1 + X_2 + X_3 + X_4 + X_5 + X_6\text{"}$$

The sum $\sum_{i=1}^{6} X_i$ is 210. We might be interested in other sums, for example,

$$\sum_{i=1}^{3} X_i = 60; \qquad \sum_{i=4}^{6} X_i = 150; \qquad \text{and} \qquad \sum_{i=2}^{5} X_i = 140$$

The "60" is obtained by starting at "X_1" and adding up through "X_3." In other words,

$$\sum_{i=1}^{3} X_i = X_1 + X_2 + X_3 = 10 + 20 + 30$$

Can you see how "150" and "140" were obtained? Find these sums:

$$\sum_{i=2}^{4} X_i = ? \qquad \sum_{i=3}^{6} X_i = ?$$

The sums are 90 and 180, respectively.

In our examples, we used the symbol combination

$$\sum_{i=a}^{b}$$

to show that we were adding some variable that has a subscript i and we were to start adding with the element having the subscript a and continue adding until we had included the element having the subscript b.

Usually (but not always), in statistics the lower subscript is 1 and the upper subscript is N. That is, usually, we sum over *all* elements in the number series. Because this is true, you might see the symbol

$$\sum_{i}$$

which means, "Sum over *all* of the numbers having the subscript i." If there is only one subscript and thus no confusion over what is to be added, even the "i" is dropped. So, returning to the example, we have equivalent formulas:

$$\sum_{i=1}^{6} X_i = 210; \qquad \sum_{i} X_i = 210; \qquad \text{or} \qquad \sum X_i = 210$$

PROBLEMS

Here is another random number series. Assign subscripts to the numbers in the order that they appear in the list. Find the indicated sums. Let's name the variable "A":

$$7, 1, 8, 3, 6, 2, 9, 4, 6, 3$$

8.9. $\displaystyle\sum_{i=1}^{N} A_i = ?$

8.10. $\displaystyle\sum A_i = ?$

8.11. $\displaystyle\sum_{i=1}^{5} A_i = ?$

8.12. $\displaystyle\sum_{i=6}^{N} A_i = ?$

8.13. $\displaystyle\sum_{i=3}^{7} A_i = ?$

SUMMING CONSTANTS

Summing over constants presents a problem:

$$\sum_{i=1}^{6} 5$$

means, literally, "Add up 5 six times." So the sum is 30! The answer is $6 \times 5 = 30$.

In general, we can write

$$\sum_{i=1}^{N} c = Nc$$

where c is a constant (which, in this case, means merely that c does not have an i subscript). So we have

$$\sum_{i=1}^{7} 4 = 28; \qquad \sum_{j=1}^{25} K = 25K; \qquad \text{and} \qquad \sum_{k=1}^{N} M = NM$$

Problems

Find the indicated sums using the following set of data.

i	1	2	3	4	5
B	20	10	5	25	10

8.14. $\sum B_i = ?$

8.15. $\sum_{i=1}^{N} 14 = ?$

8.16. $\sum_{i=1}^{N} (B_i - 10)$ $\left[Hint. \ \sum_{i=1}^{N}(B_i - 10) = \sum_{i=1}^{N} B_i - \sum_{i=1}^{N} 10 \right]$

8.17. $\sum_{i=1}^{N} B_i - 10 = ?$ (Hint. This is different from Problem 8.16.)

8.18. $\sum_{i=1}^{N} 2B_i = ?$ (Hint. This is the same as $2 \sum B_i$.)

8.19. $\sum_{i=1}^{N} (B_i - 14) = ?$

8.20. $\sum_{i=1}^{N} (B_i - 20) = ?$

Double subscripts and double summation

Suppose each of N children were measured in regard to height by two persons. We could call the height measurements "H_1" and "H_2," for measurements one and two, respectively. Now, let's call the subscript "j" and we can say $j = 1, 2$, that is, j can be 1 or 2 only. Now let's denote each of the N individuals by the subscript i:

"H_{ji}" will mean, "The jth measure of height for individual i."
"H_{27}" will mean, in particular, "The second height measurement for the seventh individual."

We can define "average height for each individual" as one-half the sum of the two measurements. Symbolically, let's call the average "A," and we can write, in general,

$$A_i = (\tfrac{1}{2})(H_{1i} + H_{2i})$$

or, in summation notation,

$$A_i = (\tfrac{1}{2})\sum_{j=1}^{2} H_{ji}.$$

We can also talk about the sum of the first height measurements for all individuals. This is

$$\sum_{i=1}^{N} H_{1i}.$$

Also, we have $\sum_{i}^{N} H_{2i}$ as the sum of the second height measurements for all individuals.

Now, we can also calculate a *double* sum, that is, the grand total of all measurements added over all individuals. This is written with a double $\sum$ as

$$\sum_{i=1}^{N} \sum_{j=1}^{2} H_{ji}$$

or equivalently,

$$\sum_{j=1}^{2} \sum_{i=1}^{N} H_{ji}$$

Notice that the order of summing is irrelevant (commutative principle of addition). However, the order of the subscripts of H is important. The first, j, refers to whether the measurement is the first one or the second one; while the second, i, refers to the individual's identification.

PROBLEMS

Here are some more problems. The first two sets provide experience with simulated statistical data while the third is a set of games that demonstrate summation principles.

8.21. Consider this data table showing the hours in which a computer has been used by several research investigators.

				Day				
Investigator	1	2	3	4	5	6	7	
1	2	1	1	2	0	1	1	*8*
2	4	0	0	3	4	1	0	*12*
3	1	3	5	0	1	2	5	*17*
4	3	1	2	4	1	3	2	*16*
5	0	0	0	5	1	1	0	*7*
								0

Let "i" be the "investigator" subscript and "j" be the "day" subscript. Let "H_{ij}" stand for "hours of computer use by investigator i on day j." Find the following.

a. H_{23} b. H_{45}

c. H_{36}

g. $\sum\limits_{i=1}^{5} H_{i1}$

d. $\sum\limits_{j=1}^{7} H_{1j}$

h. $\sum\limits_{i=1}^{5} H_{i6}$

e. $\sum\limits_{j=1}^{7} H_{2j}$

i. $\sum\limits_{j=1}^{7}\sum\limits_{i=1}^{5} H_{ij}$

f. $\sum\limits_{j=1}^{7} H_{3j}$

j. $\sum\limits_{j=3}^{6}\sum\limits_{i=2}^{5} H_{ij}$

8.22. Consider this random series of numbers:

$$4, 6, 3, 2, 7, 8$$

Assign subscripts $i = 1, \ldots, 6$ to the numbers in the order that they occur above. Call the numbers "$X_i, i = 1, \ldots, 6$."

a. Show that $\sum X_i = \sum\limits_{i=1}^{3} X_i + \sum\limits_{i=4}^{6} X_i$

b. If $M = (\sum X_i)/6$, show that $\sum (X_i - M) = 0$

c. Find $\sum (X_i - M)^2$

d. Find $\sum X_i^2 - NM^2$

e. Find $\sum\limits_{i=1}^{N}(X_i + 1) + 1$

f. Find $\sum\limits_{i=1}^{N}(X_i + 1)^2$

g. Find $\left(\sum\limits_{i=1}^{N}(X_i + 1)\right)^2$

8.23. The following problems involve no data.

a. $\sum\limits_{i=1}^{5} i = ?$

d. $\sum\limits_{i=1}^{2}\sum\limits_{j=1}^{3} ij = ?$

b. $\sum\limits_{i=1}^{4} i^2 = ?$

e. $\sum\limits_{i=1}^{5} (i + c)$

c. $\sum\limits_{i=1}^{4} i + \sum\limits_{j=2}^{4} i = ?$

f. $\sum\limits_{i=1}^{5} (i + j)$

The Product Operator, Π

The operator "Π" is used to indicate a *multiplication* of a series of numbers, just as "Σ" was used to indicate a summation of a series of numbers.

Let's define a number series as follows: $Y_1 = 2$; $Y_2 = 6$; $Y_3 = 3$; $Y_4 = 3$; and $Y_5 = 4$.

The symbol combination "$\prod_{i=1}^{5} Y_i$" means "$Y_1 \cdot Y_2 \cdot Y_3 \cdot Y_4 \cdot Y_5$" or, in the particular example of interest, $2 \cdot 6 \cdot 3 \cdot 3 \cdot 4$, or 432.

The Π-operator is used in statistics occasionally; however, it is not nearly as important as the Σ-operator.

PROBLEMS

8.24. Let $X_1 = 2$; $X_2 = 3$; $X_3 = 4$; and $X_4 = 7$.

a. $\prod_{i=1}^{4} X_i = ?$

b. $\prod_{i=2}^{4} X_i = ?$

8.25. Let $X_1 = 15$; $X_2 = 75$; and $X_3 = 0$.

What is $\prod_{j=1}^{3} X_j$?

Factorials

Another special multiplication operator is the factorial. The factorial sign "!" indicates a special repeated multiplication which is used frequently in statistical applications.

Here are several examples:

$$3! = 3 \cdot 2 \cdot 1 = 6$$
$$4! = 4 \cdot 3 \cdot 2 \cdot 1 = 24$$
$$6! = 6 \cdot 5 \cdot 4 \cdot 3 \cdot 2 \cdot 1 = 720$$

In general,

$$n! = n(n-1)(n-2) \cdots 3 \cdot 2 \cdot 1,$$

where n is an *integer*.

PROBLEMS

8.26. $2! = ?$

8.27. $5! = ?$

8.28. $7! = ?$

There are some important relationships that can be induced from patterns. First, let's combine the above examples and do some operations on them:

$$7! = 5{,}040$$
$$6! = 720 = 7!/7$$
$$5! = 120 = 6!/6$$
$$4! = 24 \ = 5!/5$$
$$3! = 6 \ \ = 4!/4$$
$$2! = 2 \ \ = 3!/3$$
$$1! = 1 \ \ = 2!/2$$
$$0! = ?$$

What does "0!" mean? If we keep the pattern in the right-hand column, it is clear that $0! = 1!/1 = 1$. Usually texts merely say, "Let's define 0! as 1," but here we see that such a definition fits nicely into our mathematical pattern.

Let's look at some more patterns:

$$7! = 5{,}040$$
$$6! = 720 = 7!/7 \qquad\qquad = (7-1)!$$
$$5! = 120 = 7!/7 \cdot 6 \qquad\quad = (7-2)!$$
$$4! = 24 \ = 7!/7 \cdot 6 \cdot 5 \qquad = (7-3)!$$
$$3! = 6 \ \ = 7!/7 \cdot 6 \cdot 5 \cdot 4 \qquad = (7-4)!$$
$$2! = 2 \ \ = 7!/7 \cdot 6 \cdot 5 \cdot 4 \cdot 3 \ \ = (7-5)!$$
$$1! = 1 \ \ = 7!/7 \cdot 6 \cdot 5 \cdot 4 \cdot 3 \cdot 2 = (7-6)!$$
$$0! = 1 \ \ = 7!/7! \qquad\qquad\quad = (7-7)!$$

Again, we see the value of defining "0!" as "1." Also notice that the calculations implied by column three are greatly simplified by canceling equal numbers before doing any computation.

Here are some more problems that show canceling:

$$\frac{7!}{3!} = \frac{7 \cdot 6 \cdot 5 \cdot 4 \cdot \cancel{3} \cdot \cancel{2} \cdot \cancel{1}}{\cancel{3} \cdot \cancel{2} \cdot \cancel{1}} = 840$$

$$\frac{6!}{4!} = \frac{6 \cdot 5 \cdot \cancel{4} \cdot \cancel{3} \cdot \cancel{2} \cdot \cancel{1}}{\cancel{4} \cdot \cancel{3} \cdot \cancel{2} \cdot \cancel{1}} = 30$$

$$\frac{5!}{4!} = \frac{5 \cdot \cancel{4} \cdot \cancel{3} \cdot \cancel{2} \cdot \cancel{1}}{\cancel{4} \cdot \cancel{3} \cdot \cancel{2} \cdot \cancel{1}} = 5$$

$$\frac{5!}{4!} = \frac{5 \cdot \cancel{4!}}{\cancel{4!}} = 5$$

It is easy to see that canceling helps considerably here. Also note in the above examples a general relationship that can be expressed

$$\frac{n!}{r!} = n(n - 1)(n - 2) \cdots (r + 1)$$

where r is some integer less than n. For example, in the problem $6!/4!$, let $n = 6$ and $r = 4$. Then $r + 1 = 5$, and the answer is $6 \cdot 5 = 30$. Also, $7!/3! = 7 \cdot 6 \cdot 5 \cdot 4$.

Problems for Review

The following table shows, for a large business, the number of employees and their corresponding average salary for each of several job categories.

Category	Number	Average Salary
0	50	$ 4,000
1	40	6,000
2	20	8,000
3	10	10,000
4	10	15,000
5	7	20,000
6	3	40,000

Let's use the category code as our subscript index, j. Let's call the elements of column two "n_j" and the elements of column three "a_j."

1. Find the following.

   a. n_1 d. a_5
   b. n_3 e. $n_1 + n_2$
   c. a_1 f. $n_6(a_6)$

2. Find N where $N = \sum_{j=0}^{6} n_j$. (What is this?)

3. Use N to obtain:

   a. n_5/N
   b. $(n_1 + n_2 + n_3)/N$

4. Find $P = \sum_{j=0}^{6} n_j a_j$. (What is this?)

5. Suppose each position receives a 10% raise. Write a formula similar to the one in Problem 4 which, when evaluated with the given data, will give the new year's total wages. Call this new total P'. Determine P' with your formula and the data.

6. Find $\prod_{i=0}^{6} n_i$.

7. Here is another set of data.

i	X_{i1}	X_{i2}	X_{i3}
1	10	40	100
2	20	45	105
3	40	40	110
4	50	35	105

   a. Find $\sum X_{i1}$; $\sum X_{i2}$; $\sum X_{i3}$
   b. Show that $10 \sum_{i=1}^{4} [(X_{i1})/10] = \sum X_{i1}$

   $$\sum_{i=1}^{4} (X_{i3} - 100) + 400 = \sum X_{i3}$$

   c. Find $\sum_{i=1}^{4} X_{i1} X_{i2}$
   d. Find $\sum_{i=1}^{4} (X_{i1} - 30)(X_{i2} - 40)$
   e. Let $M_3 = (\sum X_{i3})/4$. Find: (1) $\sum (X_{i3} - M_3)$
   $\qquad\qquad\qquad\qquad\qquad\qquad$ (2) $\sum (X_{i3} - M_3)^2$

Equations and Inequalities

Introduction

A THOROUGH knowledge of equations and their manipulation is invaluable in learning statistics. Almost all statistical formulas are expressed as equations. The statistics teacher often assumes that students have the ability to manipulate an equation into various forms, so students who cannot do so are at a disadvantage. This ability will save memorization time and can even provide extra insight into some statistical concepts.

Part of the students' confusion stems from partial recall of seldom used rules that were learned in high school. Especially confusing is the use and misuse of what was called "transposition." In this chapter we will focus on the meaning of "equality" and how this helps solve equations without using mysterious rules.

We have already defined "equation" and "inequality." This chapter will deal with both. Keep in mind the fundamental notation of an equation—it is a group of letters, numerals, and other symbols, including an equal sign, which says merely that the symbols to the left of the equal sign and the symbols to the right of the equal sign have as their referents the same number. In other words, an equation gives two different names for the same number.

The Truth Value of an Equation

An equation is a declarative sentence. It says that two symbol sets refer to the same number. It is helpful to think of the equation sentence the way you do about any "word sentence"—a sentence can be true or it can be false. So both of these two equations are perfectly legitimate:

$$7 + 2 = 6$$
$$4 + 1 = 5$$

The first equation happens to be false (a lie, so to speak). The second equation is true.

Now, suppose a term in the equation contains a letter. For example, $2X + 1 = 4$. What is the truth value of this? The truth value depends partly on the range of the variable X. If X can be *only an integer*, then the truth value is *false* for any and every X since there is *no* integer that will make "$2X + 1 = 4$" a true statement. Try some and see if you can find an integer that will make it a true statement.

If X has a range of *only negative numbers*, then there is *no* X that makes "$2X + 1 = 4$" a true statement. However, if the range of X is all numbers or all positive numbers, "$2X + 1 = 4$" *can be true* for one certain element in the range of X, namely 1.5.

Solution Sets

A basic algebra problem is finding the element or elements in the range of a variable that make an equation or inequality a true sentence. These special elements are called the "*solution set*" to the equation (or inequality). The determination of the solution set is frequently called "solving the equation" or "solving the inequality." The elements in the solution set of an equation are frequently called "*roots* of the equation."

The solution set (or root) of the example equation "$2X + 1 = 4$" is the number 1.5. This is the only member of the solution set since it is the only number that will make "$2X + 1 = 4$" a true statement. You can verify that $X = 1.5$ does make "$2X + 1 = 4$" true by replacing the "X" in the equation by "1.5" and carrying out the indicated operations.

Let's look at an inequality. Let X be a variable that has only these numbers in its range: 0, 1, 2, 3, 4, 5, 6, 7, 8, and 9. The inequality is

$$2X - 3 \leq 5$$

We can determine the solution set by trying all of the elements in the range of X (Table 9.1).

From the table it is quite clear that the solution set consists of (0, 1, 2, 3, and 4). Any of these numbers makes the inequality a true sentence. All other numbers make the inequality a false statement.

Here is another example. Let X range over all real numbers. What is the solution set to the equation

$$(X + 1)^2 = X^2 + 2X + 1$$

Table 9.1

X	2X − 3	Is 2X − 3 ≤ 5?	Truth Value
0	−3	Yes	True
1	−1	Yes	True
2	1	Yes	True
3	3	Yes	True
4	5	Yes	True
5	7	No	False
6	9	No	False
7	11	No	False
8	13	No	False
9	15	No	False

It turns out that the solution set contains *all* real numbers. That is, every possible X makes the equation true. This is a common type of equation and is called an "*identity*." Here are some common identities with which you are familiar:

$$aX = Xa$$

$$X + a = a + X$$

$$-(X + a) = -X - a$$

These equations are always true statements no matter what numbers are substituted for either variable X or variable a.

PROBLEMS

Let X have the following range:

$$(-4, -3, -2, -1, 0, 1, 2, 3, 4)$$

Find the solution set for the following equations and inequalities. You can do this by trying all numbers in the range of X.

9.1. $3X + 10 = 4$ $3X = -6$ $X = -2$

9.2. $X^2 + 1 = 10$ $X^2 = 9$ $X = 3$

9.3. $2X + 4 > 0$ $2X = -4$ $X = -2$

9.4. $3X + 12 \geq 0$

9.5. $4X + 1 = 2$

Algebraic Determination of Solution Sets

INTRODUCTION

This section will be limited to the discussion of equations having one variable of interest. These cases are commonly called "one equation with one unknown." This is a slight misnomer except in context. We might have an equation relating several variables and constants that are "unknown" in the sense that a specific numeral is not used. Here is an example:

$$aX + Y = 10$$

This is an equation with three "unknowns," which are a, X, and Y. We can solve the equation for any of these variables (i.e., find a solution set for any of a, X, or Y). This is done by temporarily treating the other two variables as known constants. So, in context, each of the symbols "a," "X," and "Y" can be considered "known" or "unknown." Perhaps this will become clearer with practice in this section.

There are a few fundamental steps used in solving equations. These are based on the definition of the equal sign. Here are several identities, true for any number. Let's start with an obviously true equation, $X = X$. This is an identity. It is true for any number. Now let "a" represent any real number. Our principles are as follows.

1. $X + a = X + a$

2. $X - a = X - a$

3. $aX = aX$

4. $\dfrac{X}{a} = \dfrac{X}{a}$

The first two identities state merely that the truth value of an equation is unchanged if the same number is added to both sides of an equation (1) or if the same number is subtracted from both sides of an equation (2). The second two identities state that the truth value of an equation is unchanged if both sides are multiplied by the same number (3) or if both sides are divided by the same number (4).

The algebraic determination of solution sets is based almost entirely on these few principles. All of them can be generalized with one basic idea—the truth value of an equation is unchanged if, for whatever modification is made on the part of the equation to the left of the equal sign, a comparable modification is also made to the part of the equation on the right of the equal sign.

The next few sections will show applications of these principles.

EQUATIONS OF THE FORM $X + a = b$

Let's do several examples using identities 1 and 2. In each of these examples, the solution set will consist of only one number. All letter symbols used will designate variables whose range consists of all real numbers.

1. $x + 5 = 10$

   Step.   Subtract 5 from each side.

$$x + 5 - 5 = 10 - 5$$

   Solution Set. $x = 5$

2. $x - 6 = 30$

   Step.   Add 6 to each side.

$$x - 6 + 6 = 30 + 6$$

   Solution Set. $x = 36$

3. $B - 10 = 0$

   Step.   Add 10 to both sides.

$$B - 10 + 10 = 0 + 10$$

   Solution Set. $B = 10$

4. $N + 5 = 25$

   Step.   Subtract 5 from each side.

$$N + 5 - 5 = 25 - 5$$

   Solution Set. $N = 20$

5. $K + a = 5$ (Find the solution set for K.)

   Step.   Subtract a from both sides.

$$K + a - a = 5 - a$$

   Solution Set. $K = 5 - a$

6. $K + a = 5$ (Find the solution set for a.)

   Step.   Subtract K from both sides.

$$K + a - K = 5 - K$$

   Solution Set. $a = 5 - K$

In Examples 5 and 6, do you see how K and a are treated as either "known" or "unknown" in the solutions?

PROBLEMS

In each of the following problems, find the solution set for Y. Consider the range of all variables to be all real numbers.

9.6. $Y + 6 = 7$ $Y = 1$

9.7. $Y + 9 = 15$ $Y = 6$

9.8. $Y - 2 = 7$ $Y = 9$

9.9. $Y - 20 = 40$ $Y = 60$

9.10. $Y + 5 - 2 + 6 = 0$ $Y = -5 + 2 - 6 = -9$

9.11. $Y + a = 0$ $Y = -a$

9.12. $Y + 3a = 2$ $Y = 2 - 3a$

9.13. $Y - 2(a + 1) = a + 1$

9.14. $17 + Y - 30 = 40$ $Y = 40 - 17 + 30 = 53$

9.15. $b + c = Y - 4$ $-Y = -4 - b - c$ $Y = 4 + b + c$

EQUATIONS OF THE FORM $aX = b$

Let's do several examples using identities 3 and 4. In each of these examples, the solution set will consist of a single number. All letters will designate variables whose range is all real numbers.

1. $5Z = 10$

    Step.   Divide both sides by 5.

$$\frac{5Z}{5} = \frac{10}{5}$$

    Solution Set. $Z = 2$

2. $\dfrac{Y}{4} = 3$

    Step.   Multiply both sides by 4.

$$\frac{Y}{4} \cdot 4 = 3 \cdot 4$$

    Solution Set. $Y = 12$

3. $.5Y = 2$

    Step.   Multiply both sides by 2.

$$2(.5)Y = 2 \cdot 2$$

    Solution Set. $Y = 4$

4.
$$\frac{3Y}{4} = 6$$

Step.   Multiply both sides by $\frac{4}{3}$.

$$\tfrac{4}{3} \cdot \tfrac{3}{4} \cdot Y = \tfrac{4}{3} \cdot 6$$

Solution Set. $Y = 8$

5. $3A = 9$

Step.   Divide both sides by 3.

$$\frac{3A}{3} = \frac{9}{3}$$

Solution Set. $A = 3$

6. $2.03B = -8.12$

Step.   Divide both sides by 2.03.

$$\frac{2.03B}{2.03} = \frac{-8.12}{2.03}$$

Solution Set. $B = -4$

7. $\dfrac{D}{a} = 7$ (Find the solution set for D.)

Step.   Multiply both sides by a.

$$\frac{aD}{a} = 7 \cdot a$$

Solution Set. $D = 7a$

Do you see how the multiplier or divisor is chosen for each example? Do you see, in the previous section, why each addend was chosen? Each equation is manipulated until the variable of interest (the "unknown") is on one side of the equal sign and the solution is on the other side of the equal sign.

PROBLEMS

Find the solution sets for these equations. The range of all variables is all real numbers.

9.16. $C/4 = \frac{1}{8}$

9.17. $5X = 5$

9.18. $2X/3 = \frac{3}{4}$

9.19. $-.7Y = .42$

9.20. $8Z = -4$

9.21. Solve these for A.

a. $aA = b$ $A = \frac{b}{a}$

b. $(a/b)A = 6$ $6\frac{b}{a}$ or $\frac{6b}{a}$

c. $A/ab = c$ $\frac{A}{ab} = c$ $A = abc$

d. $(A/a)b = c$ $\frac{A}{a} \cdot \frac{b}{1} = \frac{Ab}{a} = c$ $Ab = ac$ $A = \frac{ac}{b}$

e. $7A \div \tfrac{1}{3}b = 9$ $7A \div \frac{b}{3}$

$7A \cdot \frac{3}{b} = 9$

(handwritten margin notes:)
$\frac{a}{b}A = 6$

$\frac{Aa}{b} = 6$

$Aa = 6b$

$A = 6\frac{b}{a}$

$\frac{21A}{b} = 9$

$9b = 21A$

$21A = 9b$

$7A = 3\frac{b}{?}$

$A = \frac{3b}{?}$

EQUATIONS OF THE FORM $aX + b = c$

Let's put our rules together now.

1. $2X + 4 = 6$

Step 1.   Subtract 4 from each side.

$2X = 2$

Step 2.   Divide both sides by 2.

Solution Set. $X = 1$

You can reverse the steps, but it is harder. Let's do example 1 again.

1. $2X + 4 = 6$

Step 1.   Divide both sides by 2.

$X + 2 = 3$ (4 *must be divided also.*)

Step 2.   Subtract 2 from each side.

Solution Set. $X = 1$

It is usually easier to do the addition step first.

2. $.25B - 6.5 = -1.25$

Step 1.   Add 6.5 to both sides.

$.25B = 5.25$

Step 2.   Multiply each side by 4 (or divide by .25).

Solution Set. $B = 21$

3. $3X + 2 = 1$

Step 1.   Subtract 2 from each side.

$3X = -1$

Step 2.   Divide both sides by 3.

Solution Set. $X = -\tfrac{1}{3}$

4. $$aY + b = c \text{ (Solve for } Y.)$$

Step 1.   Subtract b from both sides.

$$aY = c - b$$

Step 2.   Divide both sides by a.

Solution Set. $$Y = \frac{c - b}{a}$$

PROBLEMS

In each of the following problems, find the solution set for x. The range of each variable is all real numbers.

9.22. $6x + 2 = 128$

9.23. $(x/7) + 1 = 0$

9.24. $.01x - 1.02 = .31$

9.25. $150 + 15x = 35$

9.26. $7 = 2x + 3$

9.27. $10 = x + 12 - 2x + 1$

9.28. $2x = 31 + x$ (*Hint.* Step 1 should be to add $-x$ to both sides.)

9.29. $2x + 10 = x + 60$

9.30. $\dfrac{2x + 1}{x + 6} = 1$

9.31. $4(x + 1) = \left[2 + 4\left(\dfrac{x}{2}\right)\right]2$

ADDING "LIKE TERMS" AND FACTORING IN AN EQUATION

The use of letter symbols in an equation introduces a trouble spot that needs to be mentioned. Numerical symbols such as "3 × 4" and "4 + 2" are easily rewritten as "12" and "6" in the process of equation solving. However, alphabetic symbols such as "$a + 6$" and "$c \cdot d$" cannot be simplified since the letters refer to either unknown constants or to any of many members of the range of variables. Frequently this causes no problem.

But let's look at some cases where problems can arise. Find the solution set for X in this equation:

$$3X + 4aX + 2bX = c$$

None of the terms can be added together. They are "*unlike terms.*" The solution in this case is to recognize that

$$3X + 4aX + 2bX = (3 + 4a + 2b)X$$

This is called "*factoring.*" The solution is

$$X = c/(3 + 4a + 2b)$$

Find the solution set for Y in this equation: $5Y + 10Y = 30$. In this case, "$5Y$" and "$10Y$" are "*like terms*" and can be added together to give $15Y = 30$, and the solution is seen to be $Y = 2$.

Solve this equation for A. $10A + 15bA + 20A + 5 = 0$. We can combine "like terms" "$10A$" and "$20A$" to give "$30A$," but additional simplification is not possible. The step-by-step solution is below:

$$10A + 15bA + 20A + 5 = 0$$
$$30A + 15bA + 5 = 0$$
$$(30 + 15b)A + 5 = 0 \text{ (factoring)}$$
$$(30 + 15b)A = -5$$
$$A = -5/(30 + 15b)$$

Find the solution set for Z in the equation

$$5Z + bZ + 7Z + 4bZ = b + 1$$

Combining like terms. $12Z + 5bZ = b + 1$

Factoring. $(12 + 5b)Z = b + 1$

Solution Set. $Z = (b + 1)/(12 + 5b)$

PROBLEMS

Find the solution sets for x in the following equations.

9.32. $15x + 25x + 2x + 10 = 4x + 3$

9.33. $5ax + 10bx + 2ax + 6bx = 10$

9.34. $7x + ax = 7 + a$

9.35. $6 + 5a + 2a + 16 + x = 2$

9.36. $2ax + 3bx + 15 = 0$

9.37. $10x + 20x = 0$

SOLVING INEQUALITIES

Inequalities are generally solved the same way that equations are solved. However, some operations do bring up a special problem in regard to the direction of the inequality sign.

Let's look at some examples that do not involve any new principles. Let Y range over all real numbers. Let's find the solution set for a few inequalities.

1. $X + 5 \leq 10$

Step.   Subtract 5 from both sides.

Solution Set. $X \leq 5$ (The solution set contains all numbers equal to or less than 5.)

2. $\qquad\qquad\qquad 6X \geq 36$

Step. $\qquad\qquad\qquad$ Divide both sides by 6.

Solution Set. $\qquad\qquad$ $X \geq 6$ (All numbers equal to or greater than 6 will make "$6X \geq 36$" a true statement.)

3. $\qquad\qquad\qquad 4X - 3 > 9$

Step 1. $\qquad\qquad\qquad$ Add 3 to both sides.

$\qquad\qquad\qquad 4X > 12$

Step 2. $\qquad\qquad\qquad$ Divide both sides by 4.

Solution Set. $\qquad\qquad$ $X > 3$ (Any number greater than 3 will make "$4X - 3 > 9$" true.)

Each of the three problems above involves no new ideas. However, here are some new problems.

4. $\qquad\qquad$ $5 - X < 4$ (This can be solved by going through two steps.)

Step 1. $\qquad\qquad\qquad$ Add X to both sides.

$\qquad\qquad\qquad 5 < 4 + X$

Step 2. $\qquad\qquad\qquad$ Subtract 4 from both sides.

Solution Set. $\qquad\qquad$ $1 < X$ (or, equivalently, $X > 1$).

However, the problem can be approached in another way in which confusion can arise.

$\qquad\qquad\qquad 5 - X < 4$

Step 1. $\qquad\qquad\qquad$ Subtract 5 from each side.

$\qquad\qquad\qquad -X < -1$

Step 2. $\qquad\qquad\qquad$ Multiply each side by -1.

This seems like the appropriate second step. The result appears to be "$X < 1$." But a quick test using $X = 2$ shows that "$-X < -1$" and "$X < 1$" are *not* equivalent. They are "$-2 < -1$" which is true and "$2 < 1$" which is false. *The truth value was changed* by multiplying both sides of the *inequality* by a negative number. The proper way to handle this is by *reversing the direction* of the inequality sign.

So our Step 2 should yield "$X > 1$" as the definition of the solution set. Think about this for awhile. In general, if $-X$ is *less* than a constant, then X must be *greater* than the *negative* of the constant:

$$\text{``}{-X} < c\text{''} \qquad \text{implies} \qquad \text{``}X > -c\text{''}$$

5. Let the range of Y be all positive numbers. Find all Y such that

$$\frac{1}{Y} < 5$$

This problem can be handled in several steps—

$$\frac{1}{Y} < 5$$

$$1 < 5Y$$

$$\frac{1}{5} < Y$$

Or, it can be handled directly by *inverting both sides* and *reversing* the inequality sign:

$$\text{``}\frac{1}{Y} < 5\text{''} \qquad \text{implies} \qquad \text{``}Y > \tfrac{1}{5}\text{''}$$

Be sure that you realize that "$\tfrac{1}{5} < Y$" and "$Y > \tfrac{1}{5}$" mean the same thing. Note that if Y could refer to negative *and* positive numbers, then all negative numbers *and* all $Y > \tfrac{1}{5}$ are included in the solution set.

6. $-17 - 3Z > 4$

    Step 1.   Add 17 to both sides.

                          $-3Z > 21$

    Step 2.   Divide both sides by -3 *and* change the direction of the inequality sign.

    Solution Set. $Z < -7$

7. $\dfrac{16}{X + 4} > 2$

    Step 1.   Invert both sides and change the direction of the inequality sign.

                          $\dfrac{X + 4}{16} < \dfrac{1}{2}$

    Step 2.   Multiply both sides by 16.

                        $x + 4 < 8$

    Step 3.   Subtract 4 from both sides.

    Solution Set. $X < 4$

                        Note that if X includes negative numbers in its range, then the solution set also includes negative numbers between -4 and 0.

PROBLEMS

Find the formulas for and the element of the solution sets for k for these inequalities if k has the following range:

$$(-8, -6, -4, -2, 0, 2, 4, 6, 8)$$

9.38. $k + 2 > 4$

9.39. $3k - 6 \leq 0$

9.40. $4 - k > 3$

9.41. $2k + 1 < 4$

9.42. $\dfrac{4}{10 - k} \leq 2$

9.43. $\dfrac{10k}{2k + 3k} \geq 2$

Constructing Equations to Describe Events

Certainly the manipulation of formulas, as you have done in the preceding sections, constitutes a large part of algebra applications in the statistics class. However, for actual research applications of mathematics, a more important application of algebra is the ability to develop formulas that describe sets of events. The researcher refers to this as "building a model." There are also problems in writing useful formulas that do not involve model building. However, let's not separate these two problems and instead think generally about describing a problem in terms of an equation of some sort.

Let's do some examples. We wish to write an equation for each of several problems.

PROBLEM 1. Each of a set of N individuals has a score on a test. Adjust each score by subtracting a constant. Divide the result by a second constant.

A solution (many solutions are possible). First, we must name the variables and constants of interest. The major variable is the test score. Let's call it "S" (for "score"). It needs to be subscripted since there are N individuals. We can write "S_i, $i = 1$, N" to indicate the various scores. The only other variable is the desired adjusted score. We can call this "A" (for "adjusted"). The constants can be called "C_1" and "C_2" for "constant one" and "constant two," respectively. The answer is

$$A_i = \frac{(S_i - C_1)}{C_2}$$

This equation, or model, expresses A_i as a function of S_i. It defines a one-to-one correspondence between the scores and the adjusted scores.

PROBLEM 2. Suppose there exists a set of numbers. The sum of these numbers divided by 25 is equal to 50. What is the sum of the numbers?

A solution. Let's call the numbers "x." Again, this is a subscripted variable. So we can write the sum using summation notation as $\sum x_i$. The problem says to divide the sum by 25 and set it equal to 50. The equation is

$$\frac{\sum x_i}{25} = 50$$

The solution set for the unknown $\sum x_i$ is 50×25, or 1250.

PROBLEM 3. The socioeconomic index is a sum of the occupational index plus three times the educational index.

A solution. $S = O + 3E$

where

S is the socioeconomic rating
O is the occupation rating
E is the education rating

PROBLEM 4. An intelligence index is a weighted sum of six variables plus a constant. The weights and constant are to be estimated from a set of data.

A solution. $I = \sum_{i=1}^{6} W_i X_i + C$

where

the W_i are six unknown weights
C is an unknown constant
X_i refers to the six variables
I is the intelligence index

Usually it helps to avoid the traditional use of X as the variable in every equation. The preferred variable symbol is one that can be easily associated with the variable. First letters are usually good choices.

PROBLEMS 5. The number μ is about 60 plus or minus 5.

A solution. $55 < \mu < 65$

or

$$55 \leq \mu \leq 65$$

This can be translated as "μ is between 55 and 65."

Solution Sets to Selected Nonlinear Equations in One Unknown

INTRODUCTION

Most (but not all) of the equations we have used so far have been of the form

$$aX + b = c$$

where a, b, and c are constants and X is a variable of interest. Equations of this form are called "linear equations" since a graph constructed from the equation will be a straight line (graphing will be discussed in a later section).

Linear equations can have more than one variable. Let X and Y be two variables and a, b, c, and d be constants. Then the general form of a linear equation with two variables is

$$aX + bY + c = d$$

This is often called a "linear function of X and Y" or a "linear composite (or 'weighted composite') of X and Y."

One general family of equations is polynomial equations. The linear equations above are special examples of polynomial equations. Other polynomial equations are characterized by exponents on the variables, such as squares, cubes, etc. Polynomials that involve an exponent of "two" are called "*quadratic* equations." We shall talk only about quadratic equations in this section.

SOLUTION SETS FOR QUADRATIC EQUATIONS

The solution set to a quadratic equation in one unknown usually contains two elements (or "roots"). Sometimes there is only one element (or root) in the solution set. Here is a simple example. For what X does $X^2 = 4$? The solution set to "$X^2 = 4$" consists of -2 and 2, since $(-2)^2 = 4$ and $2^2 = 4$. One major difficulty arises. The roots are not necessarily real numbers and are often irrational numbers. For example, the roots of $X^2 = -1$ are called complex (imaginary) numbers and the roots of $X^2 = 5$ are irrational ($-\sqrt{5}$ and $\sqrt{5}$). We will seldom have statistical examples involving complex roots, so let's ignore equations that have complex roots. However, irrational roots are as common as rational ones.

The determination of solution sets to quadratics can be difficult. The solution usually involves factoring the equation using the relationships

$$(X + a)^2 = X^2 + 2aX + a^2$$

or

$$(X + a)(X + b) = X^2 + (a + b)X + ab.$$

In fact, any quadratic can be written in the latter form. If the roots are a and b, then the equation can be written

$$(X - a)(X - b) = 0$$

Multiplying the two factors generates the quadratic equation in standard form:

$$(X - a)(X - b) = 0$$
$$X^2 - aX - bX + ab = 0$$
$$X^2 - (a + b)X + ab = 0$$

Here is an example application. Find the roots of $X^2 + 3X + 2 = 0$. This equation factors into

$$(X + 2)(X + 1) = 0$$

The roots are -2 and -1. These are the two roots because $(X + 2)(X + 1)$ will be zero if $X + 2 = 0$ *or* if $X + 1 = 0$. You can substitute -2 and -1 into $X^2 + 3X + 2 = 0$ and see if *both* lead to a true statement.

It is not always easy to factor the equation. The above example is fairly easy. We want to find numbers a and b such that

$$(X + a)(X + b) = X^2 + (a + b)X + ab$$

where $a + b = 3$ and $ab = 2$. Obviously, 2 and 1 will do nicely. Try these.

1. $X^2 + 4X + 4 = 0$ $(x+2)(x+2)$
2. $X^2 + 5X + 6 = 0$ $(x+2)(x+3)$
3. $X^2 + 2X - 3 = 0$ $(x-1)(x+3)$

The factors are: (1) $(X + 2)(X + 2)$ since $2 + 2 = 4$ and $2 \cdot 2 = 4$
                      (2) $(X + 2)(X + 3)$ since $2 + 3 = 5$ and $2 \cdot 3 = 6$
                      (3) $(X - 1)(X + 3)$ since $-1 + 3 = 2$ and $-1 \cdot 3 = -3$

The roots in each of these examples are:

(1) -2
(2) -2 and -3
(3) 1 and -3

If the equation can be written in general form, and it should always be possible to do so, then the roots can be determined by formulas. If the equation is

$$aX^2 + bX + c = 0$$

then the two roots are X_1 and X_2, where

$$X_1 = \frac{-b + \sqrt{b^2 - 4ac}}{2a}$$

and

$$X_2 = \frac{-b - \sqrt{b^2 - 4ac}}{2a}$$

In the first example above, we have $a = 1$, $b = 4$, and $c = 4$. The roots are

$$X_1 = \frac{-4 + \sqrt{16 - 16}}{2} = -2$$

and

$$X_2 = \frac{(-4 + 0)}{2} = -2$$

Work out the other two examples by the formula.

PROBLEMS

Solve these quadratics. Use the factor method or the formula.

9.44. $X^2 - 5X + 4 = 0$

9.45. $2X^2 + 3X + 1 = 0$

9.46. $9X^2 + 12X + 4 = 0$

9.47. $X^2 - 5 = 0$

9.48. $6X^2 + X = 1$

Problems Involving Special Statistical Applications

This section uses common statistical equations and symbols. The symbols and equations are not explained. Treat this purely as an academic exercise using content that is highly relevant to your future statistical studies.

1. Find all N such that

$$\frac{d}{10/\sqrt{N}} \geq 2$$

2. Solve

$$\sigma = \sqrt{\frac{n \sum x^2 - (\sum x)^2}{n(n - 1)}}$$

for $\sum x^2$.

3. What values of $\bar{X}$ make $\dfrac{\bar{X} - 100}{25} > 2$ and $\dfrac{\bar{X} - 100}{25} < -2$?

4. Suppose we incorrectly calculate $\left(\sum_{i=1}^{10} X_i\right)\Big/10$ to be 50. Suppose our error was thinking that $X_1 = 5$ when in fact $X_1 = 10$. What is the *correct* value of $\left(\sum_{i=1}^{10} X_i\right)\Big/10$?

5. Suppose $s = 10$ was calculated by the formula $s = \sqrt{\dfrac{\sum x^2}{11}}$. It should have been calculated using $s' = \sqrt{\dfrac{\sum x^2}{10}}$. Find the correct value s'. (*Hint.* Solve s for $\sum x^2$.)

6. Solve for r':

$$r = (r'y)/\sqrt{pq}$$

7. If $0 \leq q \leq 1$ and $0 \leq p \leq 1$ and if $p + q = 1$, what is the maximum value that pq can have? What is the minimum? Express the range of pq in an inequality.

8. Find N if $C = \sqrt{\dfrac{x}{N + x}}$.

9. Find $\sum X_i Y_i$ in the equation

$$\sum x_i y_i = \frac{N \sum X_i Y_i - \sum X_i \sum Y_i}{N}$$

if $\sum x_i y_i = 20$, $\sum X_i = 15$, $\sum Y_i = 20$, and $N = 30$.

10. Suppose $s = \sqrt{\dfrac{x}{n}}$ and $s' = \sqrt{\dfrac{x}{n-1}}$. Let r be the ratio of s to s'. Express r as a function of n.

Systems of Equations

Introduction

THE examination of systems of equations is a very common mathematical and statistical problem. The phrase "system of equations" refers to a set of equations having the same unknown parameters. For example, what are a and b if it is known that

$$a + b = 1 \quad and \quad 2a + b = 2$$

The two equations are a "system of equations." The question posed is the usual question asked about such systems—namely, for what numbers a and b are *both* equations simultaneously true statements? In the example, it turns out that if $a = 1$ and $b = 0$, then both equations are true statements. The major content of this chapter consists of the methods for determining a and b. This process commonly is called "the simultaneous solution of a system of equations."

The most common example of a statistical application is prediction. The statistician usually considers prediction to be a major goal of empirical research, so the importance of the techniques are obvious. The techniques discussed are also helpful in understanding relationships among related statistical formulas.

Simultaneous Solutions to Two Equations with Two Unknowns

SUBSTITUTION

Let's solve the system already given as an example. We want to find the solution set for a and b that will make these two equations true simultaneously:

$$a + b = 1 \quad and \quad 2a + b = 2$$

We can make several substitutions. The first equation can be rewritten as either "$a = 1 - b$" or "$b = 1 - a$." We can use either of these in equation two. Here is one solution:

$$2a + b = 2$$
$$2(1 - b) + b = 2 \text{ (substituting "} 1 - b \text{" for "} a \text{")}$$
$$2 - 2b + b = 2$$
$$2 - b = 2$$
$$-b = 0$$
$$b = 0$$

Now we can use the fact that $b = 0$ in the original equation $a + b = 1$, to get $a + 0 = 1$, or $a = 1$. The solution set is $a = 1$ *and* $b = 0$.

On the other hand, $a + b = 1$ can be solved for b, namely, $b = 1 - a$, and this can be used in equation two:

$$2a + b = 2$$
$$2a + (1 - a) = 2 \text{ (substituting "} 1 - a \text{" for "} b \text{")}$$
$$a + 1 = 2$$
$$a = 1$$

The value $a = 1$ can be substituted into $a + b = 1$ to give $1 + b = 1$, or $b = 0$.

There are two other ways to solve these equations by substitution. We can start by solving equation two for either a or b, and use the result in equation one. From equation two, we have $a = (2 - b)/2$ or $b = 2 - 2a$. When substituted into equation one, $a + b = 1$, the first result gives $b = 0$ as shown below:

$$\frac{2 - b}{2} + b = 1 \left(\text{substituting } \frac{\text{"} 2 - b \text{"}}{2} \text{ for "} a \text{"} \right)$$
$$1 - b/2 + b = 1$$
$$\frac{b}{2} = 0$$
$$b = 0$$

The second result, $b = 2 - 2a$, gives $a = 1$ as shown below:

$$a + (2 - 2a) = 1 \text{ (substituting "} 2 - 2a \text{" for "} b \text{")}$$
$$-a + 2 = 1$$
$$-a = -1$$
$$a = 1$$

So we see that there are several approaches to solving the equations by substitution. The choice is entirely one of convenience—use the substitutions that appear simplest.

Here is another example. Find x and y such that (1) $3x + 2y = 6$ and (2) $4x + y = 10$. The "easiest" substitution appears to be $y = 10 - 4x$, from equation two. Used in equation one, we have

$$3x + 2(10 - 4x) = 6$$
$$3x + 20 - 8x = 6$$
$$-5x = -14$$
$$x = \tfrac{14}{5} = 2.8$$

The value $x = 2.8$ can be used in either equation to find y. From equation two we have

$$y = 10 - 4x$$
$$= 10 - 4(2.8)$$
$$= 10 - 11.2$$
$$= -1.2$$

Our answers are $x = 2.8$ and $y = -1.2$

These answers can be "*checked*" by inserting them back into the equations to see if the truth values of the equations are maintained:

$$3x + 2y = 6 \qquad\qquad 4x + y = 10$$
$$3(2.8) + 2(-1.2) = 6 \qquad\qquad 4(2.8) - 1.2 = 10$$
$$8.4 - 2.4 = 6 \qquad\qquad 11.2 - 1.2 = 10$$

The substitution of the derived solution set does make both equations true statements, so the solution set is "correct."

Row operations

A second general solution deals with the addition and subtraction of the equations, usually after multiplying one equation by a constant. These operations are called "elementary row operations," a term that will have more meaning after studying the matrix section. Let's repeat the two examples. The first problem

$$2a + b = 2$$
$$a + b = 1$$

can be attacked by subtracting, term by term, the second equation from the first. The result is $(2a - a) + (b - b) = (2 - 1)$, or $a = 1$. This result can be substituted into $a + b = 1$ to give $b = 0$. This procedure consists of subtracting the same number from each side of the equal sign of equation one, since "$a + b$" and "1" are equal, according to equation two.

The second example is slightly more difficult. Solve simultaneously the following equations:

$$3x + 2y = 6$$
$$4x + y = 10$$

Let's start by multiplying the second equation by 2. This makes the coefficients of y equal to 2 in both equations, so that the y terms drop out in the subtraction of the new equation two from equation one:

$$\begin{array}{r} 3x + 2y = 6 \\ -(8x + 2y = 20) \\ \hline -5x = -14 \end{array}$$

You must keep the signs straight. The result gives $x = 2.8$. Substitution into either original equation gives $y = -1.2$. You might check these answers for accuracy as we did before by substituting both into the original equations.

Graphing

Often, systems can be solved easily by graphing. This topic will not come up until later in the text, but it is important to mention now. The two equations can be graphed. The intersection point gives the simultaneous solution to the equations.

Matrix solutions

A fourth technique for simultaneously solving equations is by matrix algebra. This is the topic of Part Three of this book. The matrix solution to equation systems is at the end of the matrix section.

Problems

10.1. Find, to three decimals, the values of b_1 and b_2 that satisfy both of these equations:

$$b_1 + .6b_2 = .7$$
$$.6b_1 + b_2 = .8$$

From equation one, we get $b_1 = .7 - .6b_2$. This can be used in the second equation to get $.6(.7 - .6b_2) + b_2 = .8$. Solve this equation for b_2 and then find b_1. Check the results for consistency.

10.2. Find the solutions to these sets of equations by row operations:

a. $2A + 5B = 0$ b. $5X + 2Y = 6$

$A + 3B = 2$ $2X + 4Y = 4$

Hints. (a) Multiply row two by 2 and remove A by subtraction.

(b) Multiply the second equation by .5 and remove Y, or multiply by 2.5 and remove X.

Simultaneous Solutions for Three Equations in Three Unknowns

Each of the solutions for two-equation systems is appropriate for three-equation systems. However, the calculations become more difficult and problems of calculation errors more prevalent. In fact, the graphic solution is virtually useless since graphs would be in three dimensions. Conceptually, however, it is still a proper solution technique.

SUBSTITUTION

Consider this example. Find X, Y, and Z, such that each of the following conditions are simultaneously true:

$$X + Y + Z = 1$$
$$X - Y + 2Z = 3$$
$$2X + Y - Z = 2$$

We can start substituting in many different ways. Take the third equation, as an example, and solve for Z. We get $Z = 2X + Y - 2$. We can substitute this result into equation one to get an equation in only X and Y. We have

$$X + Y + (2X + Y - 2) = 1$$

which we can solve for X. We obtain

$$3X + 2Y - 2 = 1 \quad \text{or} \quad 3X + 2Y = 3$$

We can also use our value for Z to reduce equation two to include X and Y only. We get

$$X - Y + 2Z = 3$$
$$X - Y + 2(2X + Y - 2) = 3$$
$$X - Y + 4X + 2Y - 4 = 3$$
$$5X + Y = 7$$

Now we have two equations in only two unknowns, which we can solve as we did before:

$$3X + 2Y = 3$$
$$5X + Y = 7$$

$Y = 7 - 5X$, so $3X + 2(7 - 5X) = 3$. We get $-7X = -11$, or $X = \frac{11}{7}$. With our value of X, we can find that $Y = -\frac{6}{7}$ and $Z = \frac{2}{7}$. You might want to check these for consistency by substituting them back into the original equations.

ROW OPERATIONS

Let's redo the same problem in another way:

$$X + Y + Z = 1$$
$$X - Y + 2Z = 3$$
$$2X + Y - Z = 2$$

We can add equation two to equation one to get

$$2X + 3Z = 4$$

and add equation two to equation three to get

$$3X + Z = 5$$

These two equations can be solved simultaneously.

$$2X + 3Z = 4$$
$$3X + Z = 5$$

Let's multiply the bottom row by 3 and subtract it from the row above it. We get

$$-7X = -11 \quad \text{or} \quad X = \frac{11}{7}$$

as before. This technique is seen to be considerably neater and quicker, at least in this example. In general, it is easier to keep track of the work using this method.

PROBLEM 10.3

Find A, B, and C, such that all of these conditions are true:

$$A + B - C = 10$$
$$2A - B + 2C = 1$$
$$3A - 2B - C = 17$$

Generalizations

It should be apparent that the techniques are applicable to larger systems of equations. It should also be apparent that the calculations can become extremely burdensome; thus simpler solutions are required. Fortunately, we not only have calculators and computers to help, but also highly systematized hand solutions are given in most statistics texts.

The general condition necessary, but not sufficient, for a solution is that there exist at least one equation for each unknown. That is, if there are six unknowns to find, you must have at least six equations. For example, if I asked, "For what values of X and Y is '$X + Y = 10$' true?" it should be obvious that many solutions exist, including 1 and 9, -2 and 12, 5 and 5, 6.4 and 3.6, etc. So, one more equation relating X and Y needs to be specified if a *specific* X and Y pair are to be estimated. If you have one equation and two unknowns, there is virtually too much unknown in the system.

On the other hand, we can have too many restrictions (or equations) as well, so that there are no solutions. For example, find X and Y such that

$$X + Y = 1$$
$$X + 2Y = 3$$
$$X + 3Y = 2$$

There is no such pair of X, Y values since the restrictions (equations) conflict.

Problems for Review

Solve these systems of equations:

1. $3A_1 + 2A_2 = 1$
   $2A_1 + 4A_2 = 0$

2. $.5X + 1.5Y + 2 = 0$
   $.2X + 2Y - 2 = 0$

3. $R + 2S + 2T - 11 = 0$
   $2T - R - 2S - 1 = 0$
   $T + 4R - 7 = 0$

4. $A + 3B + 5C = 1.5$
   $6A - 2B + 6C = 17$
   $2A + 4B - 8C = -4$

Part Three

BASIC MATRIX ALGEBRA

THIS part provides an introduction to the use of vector and matrix algebra as they are used in statistical applications. The notation is all-important, allowing masses of data to be described in very few symbols. Matrix notation also is conveniently used in computer programming, which accounts a great deal for its increasing popularity. Many applied statistical procedures can be discussed more easily and completely in matrix language than is possible without it.

Only the most fundamental operations will be discussed. The important goal is to become comfortable using matrix language and to learn some of the basic operations on matrices. This part has three chapters. The first introduces matrix terminology; the second reviews some elements of vector algebra; and the third focuses on matrix algebra.

Basic Matrix Concepts

Introduction

THIS chapter introduces matrix language and relates the matrix concepts to statistical applications. The fundamental value of matrix notation is the representation of large sets of data by single symbols. We will start by defining "matrix" and introduce "vector" and "scalars" as special cases.

Definitions and Symbols

MATRICES

A matrix is any set of numbers that is arranged in a rectangular pattern of rows and columns. This is the standard mathematical definition and it fits statistical data perfectly. A "data matrix" is thus a rectangular array of data.

Let's take a simple academic example first and then introduce statistical examples. Here is a set of numbers arranged in a rectangular pattern of rows and columns. The large parentheses are used to show that it is a matrix.

$$\begin{pmatrix} 7 & 6 & 5 & 4 \\ 3 & 1 & 2 & 1 \\ 1 & 3 & 0 & 4 \end{pmatrix}$$

There are three rows and four columns. The number of rows and columns is called the "order" or "dimensions" of the matrix. The order (or dimensions) of the example matrix is 3 by 4. The matrix is referred to as a "three-by-four matrix." The number of rows is always given prior to the number of columns. Here is a 3 × 2 matrix.

$$\begin{pmatrix} 7 & 1 \\ 2 & 8 \\ 3 & 0 \end{pmatrix}$$

If the number of rows equals the number of columns, the matrix is called a "*square* matrix." Here is a 3 × 3 square matrix.

$$\begin{pmatrix} 2 & 1 & 4 \\ 3 & 0 & 1 \\ 2 & 1 & 5 \end{pmatrix}$$

If a square matrix has only ones on the upper-left-to-lower-right diagonal, and zeros everywhere else, it is called an "*identity* matrix."

Here is a 4 × 4 identity matrix.

$$\begin{pmatrix} 1 & 0 & 0 & 0 \\ 0 & 1 & 0 & 0 \\ 0 & 0 & 1 & 0 \\ 0 & 0 & 0 & 1 \end{pmatrix}$$

An identity matrix is a special case of a *diagonal* matrix. A diagonal matrix is any one that has zero's everywhere except on the upper-left-to-lower-right diagonal. Here is an example of a 3 × 3 diagonal matrix that is not an identity matrix.

$$\begin{pmatrix} 5 & 0 & 0 \\ 0 & 1 & 0 \\ 0 & 0 & 2 \end{pmatrix}$$

It is customary to denote a matrix by capital letters. Some texts use only boldface capital letters; however, we shall use the usual matrix notation in italics. The numbers in the matrix are called "elements of the matrix." It is customary to use small-case subscripted letters to indicate the elements. Thus we might define a 2 × 3 matrix A with elements a_{ij}, $i = 1, 2$ and $j = 1, 2, 3$, as follows:

$$A = \begin{pmatrix} a_{11} & a_{12} & a_{13} \\ a_{21} & a_{22} & a_{23} \end{pmatrix}$$

Notice that the first subscript "i" shows the row and the second subscript "j" shows the column of each a_{ij}. Now suppose that $a_{11} = 1$, $a_{12} = 2$, $a_{13} = 5$, $a_{21} = 1$, $a_{22} = 4$, and $a_{23} = 6$.

The matrix A is

$$\begin{pmatrix} 1 & 2 & 5 \\ 1 & 4 & 6 \end{pmatrix}$$

It is obviously easy to communicate by using "A" to mean all of the six numbers (or elements) in precisely the given locations.

PROBLEMS

Let N be the matrix below.

$$\begin{pmatrix} 5 & 4 & 6 & 1 \\ 2 & 3 & 4 & 7 \\ 4 & 1 & 2 & 6 \end{pmatrix}$$

11.1. What is the order of N?

11.2. What is:

(a) n_{11}? (c) n_{31}?

(b) n_{13}? (d) n_{24}?

Here is a general matrix the order of which is $r \times c$. Let's call it X.

$$\begin{pmatrix}
x_{11} & x_{12} & x_{13} & \cdots & x_{1j} & \cdots & x_{1c} \\
x_{21} & x_{22} & x_{23} & \cdots & x_{2j} & \cdots & x_{2c} \\
x_{31} & x_{32} & x_{33} & \cdots & x_{3j} & \cdots & x_{3c} \\
\vdots & \vdots & \vdots & \cdots & \vdots & \cdots & \vdots \\
x_{i1} & x_{i2} & x_{i3} & \cdots & x_{ij} & \cdots & x_{ic} \\
\vdots & \vdots & \vdots & \cdots & \vdots & \cdots & \vdots \\
x_{r1} & x_{r2} & x_{r3} & \cdots & x_{rj} & \cdots & x_{rc}
\end{pmatrix}$$

The sets of triple dots represent continuations of the patterns. Study the patterns to be sure you can see how the subscripts change in the different rows and columns.

Another common type of matrix that is seen in statistical reports is a *symmetric* matrix. This is a matrix in which the off-diagonal elements (numbers) are related in pairs. The (i, j) element is equal to the (j, i) element. Here is an example of a 4 × 4 symmetric matrix.

$$B = \begin{pmatrix} 1 & 2 & 4 & 3 \\ 2 & 4 & 2 & 5 \\ 4 & 2 & 1 & 1 \\ 3 & 5 & 1 & 3 \end{pmatrix}$$

Note that $b_{12} = b_{21} = 2$; $b_{13} = b_{31} = 4$; and in general, $b_{ij} = b_{ji}$. Data matrices that are symmetric are usually reported by showing only one-half of the entries, since the complete table is redundant. Such matrices are common in problems of finding relationships among variables.

Vectors

Vectors can be defined most easily as matrices that have one as one dimension and a number greater than one as a second dimension. A vector is either a *row of numbers* or a *column of numbers*. A row of numbers is called a "*row vector*." Here is a *row vector* of order 4.

$$(7, 3, 2, 1)$$

A column of numbers is called a "*column vector*." Here is a *column vector* of order 3.

$$\begin{pmatrix} 2 \\ 7 \\ 8 \end{pmatrix}$$

Both of these could be considered "degenerate matrices" or matrices of order 1 × 4 and 3 × 1, respectively. However, it is convenient to consider vectors not only as special matrices, but at times it is also helpful to think of matrices as being constructed from vectors. For example, consider two row vectors of order 4.

$$(1, 2, 4, 3)$$
$$(7, 1, 2, 5)$$

These can be used together to define a 2 × 4 matrix.

$$\begin{pmatrix} 1 & 2 & 4 & 3 \\ 7 & 1 & 2 & 5 \end{pmatrix}$$

Column vectors can be used in the same way to construct matrices.

You might think that our definition of vectors conflicts with the high school physics class classical definition of a vector as being some sort of quantity that has a direction. Actually, there is no conflict, as all physical vector problems are handled quite easily with the general definition that we use in statistics. Our definition has the important property of being directly applicable to sets of data.

SCALARS

A scalar is merely a single number. It can be thought of as a "degenerate matrix" of order 1×1. However, a distinction must be made between matrices, vectors, and scalars since the failure to do so can complicate reading matrix equations. Scalars are usually symbolized by small-case letters. They can be either constants or variables, depending on context.

Up to this point, everything we have discussed can be considered as "scalar arithmetic" and "scalar algebra," since we have worked only with single numbers—not with vectors and matrices.

Statistical Applications and Data Handling

Let's construct a common statistical problem. Let's suppose the following information is taken from the personnel files of a business firm.

ID	Name	Sex	Age	Years on Job	Year's Sales Gross
1	Jones	Male	25	5	$100,000
2	Smith	Male	35	4	150,000
3	Taylor	Female	40	18	200,000
4	Adams	Female	25	2	50,000
5	Wright	Male	55	15	150,000

Let's code the data by using the ID number for names and let "1" mean "male" and let "2" mean "female." We can even ignore the identification numbers if we keep everything in order.

We can define several meaningful vectors. For example, each row constitutes a "person vector" or "data vector" for the individuals. Jones' data vector is

$$(1, 25, 5, 100,000)$$

Smith's data vector is

$$(1, 35, 4, 150{,}000)$$

On the other hand, the columns are also meaningful column vectors. They constitute "variable vectors," since they correspond to the variables under study. The "sex vector" is

$$\begin{pmatrix} 1 \\ 1 \\ 2 \\ 2 \\ 1 \end{pmatrix}$$

The "age vector" is

$$\begin{pmatrix} 25 \\ 35 \\ 40 \\ 25 \\ 55 \end{pmatrix}$$

Obviously, you can construct three more person data vectors and two more variable data vectors. The entire data matrix can be written as

$$\begin{pmatrix} 1 & 25 & 5 & 100{,}000 \\ 1 & 35 & 4 & 150{,}000 \\ 2 & 40 & 18 & 200{,}000 \\ 2 & 25 & 2 & 50{,}000 \\ 1 & 55 & 15 & 150{,}000 \end{pmatrix}$$

Now, let's call the data matrix "X." Its order is 5×4. This means, in this case, that X has 5 person-rows and 4 variable-columns.

PROBLEM 11.3

What are the *meanings* of these symbols, in the context of the above example?

a. x_{24}

b. x_{42}

c. x_{3j}

d. x_{i4}

e. $\displaystyle\sum_{i=1}^{5} x_{i4}$

The example shows one of the more common ways of organizing statistical data. It is quite common to lay out a set of data in this way and talk about "the data matrix" or "Jones' data vector," or "the vector of years-on-the-job." In fact, it is common to record the data on computer cards or tape by person vectors. Usually one card is punched for each person vector (rows). Thus, the matrix X could be a listing of the five data cards required if the data are recorded on cards, one card per person.

Problems for Review

Be sure you know the meanings of these terms.

1. Matrix.
2. Vector.
3. Scalar.
4. Row vector.
5. Column vector.
6. Square matrix.
7. Diagonal matrix.
8. Identity matrix.
9. Symmetric matrix.
10. The order of a matrix.
11. The order of a vector.

Vector Algebra

Introduction

A VECTOR is a row of numbers *or* a column of numbers. It is an extremely useful concept in many research areas, and you are probably familiar with the use of vectors in the physical sciences. The most common statistical examples are the vector of observations on several variables for one individual subject or experimental unit and the vector of observations of one variable across several subjects or experimental units. We used examples of these in the previous chapter.

There are various reasons for learning vector algebra fundamentals. One reason is because vectors are widely used in statistics application. A more urgent reason is that these concepts are useful in matrix algebra and, if you learn them before you progress to the next chapter, the understanding of matrix concepts will be simplified.

Geometric Interpretation

Let's first consider a simple physics problem. Suppose two men are dragging a large object. One man exerts a 40 pound pull, and the second man exerts a 30 pound pull. The two men are pulling at a 90° angle to each other. The actual directions of pull are arbitrary. We could say that the stronger man is pulling eastward while the less strong man is pulling northward. The basic question is to determine the direction in which the object is moving. In this case, the traditional use of the word "vector" is appropriate—a vector is a "force acting in a specific direction." Traditionally, vectors are represented by directed line segments, the length of which designates the amount of force. We can diagram the problem with two lines, one 40 units

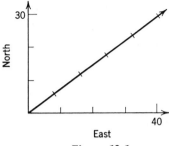

Figure 12.1

long and the other 30 units long, both pointing in appropriate directions (Figure 12.1). The diagonal line shows the direction in which the object will move, and its length (50 units) shows the force acting in the direction of movement. Each of the three lines represents force vectors. The vectors can be mathematically described, in addition to the line drawings, by the force magnitude and the direction of each vector.

However, it is also possible to describe each vector as an ordered pair of numbers written in a row or column, which is the matrix notation for a vector. We might name one vector "(40, 0)" meaning the force is 40 units directed in the eastern direction and 0 units directed in the northern direction. The other vector can be called "(0, 30)" meaning 0 units east and 30 units north. The diagonal vector (40, 30) is 40 units east and 30 units north. The order is quite important. [What is the difference between (0, 30) and (30, 0)?]

We might also consider an airplane in flight. The plane might move in any direction—north, south, etc.,—but it can also move up and down in altitude. We need three dimensions to describe its flight, so we need an ordered triad of numbers. One of the three numbers could represent the force in the north (or south) direction, the second number could show the force in the east (or west) direction, while the third number could show the force of altitude gain or loss.

It is easy enough to generalize to more dimensions. Remember the number of elements in the vector is its "dimension" or "order." The statistician usually works in many dimensions (multidimensions), one dimension for each variable of interest; or possibly one dimension for each subject in the study.

The geometric figures are awkward or even useless for describing complex vector systems of several dimensions. Fortunately, the geometric interpretation is, more often than not, unnecessary. The definition of a vector as merely a row or column of numbers makes the "direction of force" definition unnecessary.

Vector Addition and Subtraction

Let's suppose that there exists a grocer who sells only apples, oranges, and pears. He wants to record his sales. Since he has only three products, he decides to always record his sales in the same order: apples, oranges, and pears. Mrs. Jones buys 6 apples, 12 oranges, and 6 pears; however, with only three products for sale the grocer has memorized the names of his fruit and the order in which he wishes to record the sale. So he does not need to use the words "apples," "oranges," and "pears." He can record for Mrs. Jones only the number triad (6, 12, 6) and he knows exactly what this means. The next customer, Mrs. Smith, buys 12 apples, no oranges, and 24 pears. Mrs. Smith's vector is (12, 0, 24). A third customer, Mr. White, buys nothing. In fact, Mr. White returns 6 bad pears that he purchased the day before. Mr. White's vector is (0, 0, −6). No more customers arrive on this particularly bad business day. The grocer now wants to determine the day's business. Mathematically, he wants to *find the sum of the several vectors*. The solution will be a *vector of sums*. We have the following record:

$$
\begin{array}{ll}
\text{Jones} & (6, 12, 6) \\
\text{Smith} & (12, 0, 24) \\
\text{White} & (0, 0, -6)
\end{array}
$$

The grocer sold 18 apples, 12 oranges, and 30 pears. Six pears were returned. So his vector of sums is (18, 12, 24). Notice that this sum vector is obtained in the following way:

$$(6, 12, 6) + (12, 0, 24) + (0, 0, -6)$$

$$= (6 + 12 + 0, 12 + 0 + 0, 6 + 24 - 6)$$

Do you see the element-by-element addition?

In general, if a is the vector $(a_1, a_2, \ldots, a_n)$ and b is the vector $(b_1, b_2, \ldots, b_n)$, then $(a + b)$ is a vector with elements $(a_1 + b_1)$, $(a_2 + b_2)$, $(a_3 + b_3), \ldots, (a_n + b_n)$ and $(a - b)$ is a vector with elements $(a_1 - b_1)$, $(a_2 - b_2)$, $(a_3 - b_3), \ldots, (a_n - b_n)$.

Vector addition or subtraction is defined only if the vectors to be added or subtracted have the same order. For example,

$$(6, 9, 12, 17) - (2, 8, 14, 5) = (4, 1, -2, 12)$$

but $(6, 9, 12) - (2, 8, 14, 5)$ is an undefined set of symbols.

Column addition and subtraction are done the same way. For example,

$$\begin{pmatrix} 2 \\ 1 \\ 3 \end{pmatrix} + \begin{pmatrix} 1 \\ 2 \\ 4 \end{pmatrix} = \begin{pmatrix} 3 \\ 3 \\ 7 \end{pmatrix}$$

$$\begin{pmatrix} -1 \\ 4 \\ 2 \end{pmatrix} - \begin{pmatrix} 3 \\ 1 \\ -1 \end{pmatrix} = \begin{pmatrix} -4 \\ 3 \\ 3 \end{pmatrix}$$

The sum of a column vector and a row vector is undefined, even if the vectors have the same order.

PROBLEMS

Consider these vectors:

$$a = (1, 4, 2, -1) \qquad f = (1, -4, -4)$$
$$b = (3, 2, 1) \qquad g = (6, 2, 1, 3)$$
$$c = (1, 2, 4) \qquad h = (4, 2, 1)$$

$$d = \begin{pmatrix} 1 \\ 2 \\ 3 \end{pmatrix} \qquad i = \begin{pmatrix} 5 \\ -1 \\ -2 \end{pmatrix}$$

$$e = (4, 1, 2)$$

12.1. Determine which of the following operations are possible and carry out the possible additions and subtractions.

a. $b + f$

b. $b + d$

c. $d + i$

d. $b + e + h$

e. $a - b$

f. $g - a$

12.2. Show that the commutative principle holds for vector addition in the following examples.

a. $b + f = f + b$

b. $d + i = i + d$

12.3. Show that the associative principle holds for vector addition in the following example:

$$(b + c) + f = b + (c + f)$$

12.4. If u is a vector with elements $(u_1, u_2, \ldots, u_n)$ and v is a vector with elements $(v_1, v_2, \ldots, v_n)$, what is the 6th element of $u + v$? What is the ith element of $u + v$?

The Transpose Operation

The conversion of row vectors into column vectors and column vectors into row vectors is called "transposing." Let V be the vector $(1, 4, 3, 6)$. Then the vector obtained by transposing V is

$$\begin{pmatrix} 1 \\ 4 \\ 3 \\ 6 \end{pmatrix}$$

The column vector is called the "transpose of V" and is identified by a prime (') on V. That is,

$$V = (1, 4, 3, 6)$$

and

$$V' = \begin{pmatrix} 1 \\ 4 \\ 3 \\ 6 \end{pmatrix}$$

Let a be $(1, 2, 4)$ and b be $\begin{pmatrix} 3 \\ 1 \\ 2 \end{pmatrix}$. Then $(a + b)$ is undefined, but $(a + b')$ is

$$(1, 2, 4) + (3, 1, 2) = (4, 3, 6)$$

and $(a' + b)$ is

$$\begin{pmatrix} 1 \\ 2 \\ 4 \end{pmatrix} + \begin{pmatrix} 3 \\ 1 \\ 2 \end{pmatrix} = \begin{pmatrix} 4 \\ 3 \\ 6 \end{pmatrix}$$

The prime always identifies a transpose in vector and matrix symbolization.

One convenient use of transposition is for writing convenience rather than computational convenience. Column vectors can always be written in transposed form as row vectors, which is much more convenient than column form.

PROBLEMS

Let $a = (4, 2, 1, 6)$

$b = (1, 3, 2, 1)$

$$c = \begin{pmatrix} 2 \\ -3 \\ 4 \\ 1 \end{pmatrix}$$

12.5. Find the following.

a. a'

c. $b' + c$

b. c'

d. $a + c'$

12.6. If v is a vector of elements $(v_1, v_2, \ldots, v_k)$, what is the fourth element of v'? What is the ith element of v'?

Vector Multiplication

THE MULTIPLICATION OF A VECTOR BY A SCALAR

Suppose the Jones family consists of 4 persons and the Smith family consists of 8 persons. Mrs. Jones has called a store to buy food for only the Jones family. The grocer has the following list.

bread	1 loaf
meat	2 pounds
potatoes	4
lettuce	1 head
ice cream	1 quart

The Smiths drop in unexpectedly for dinner. Mrs. Jones calls the store and says "triple the order." The new list is:

bread	3 loaves
meat	6 pounds
potatoes	12
lettuce	3 heads
ice cream	3 quarts

In this example, we have an original vector of amounts and the scalar multiplier 3. The final vector resulted from multiplying the original amount vector, item-by-item, by the scalar 3.

In general, if u is the vector $(u_1, u_2, \ldots, u_k)$ and c is a scalar, then

$$cu = (cu_1, cu_2, cu_3, \ldots, cu_k)$$

PROBLEM 12.7

$$a = (1, 2, 4, 6) \qquad b = (0, -1, 2, -2)$$

Find the following:

a. $2a$
b. $3b$
c. $a - 3b$
d. $2a - 3b$
e. $3a + 2b$
f. $a' + 3b'$

PRE- AND POSTMULTIPLIER

Products of two vectors and products of two matrices do *not* commute. That is, if a and b are vectors *or* matrices, then, in general, $a \times b \neq b \times a$. Because of this, it is difficult to talk about the factors in a vector or matrix product without specifying which factor is meant.

The words "premultiplier" and "postmultiplier" are used to make this distinction. In the product "$a \times b$," it is common to say that "a is the premultiplier" and that "b is the postmultiplier." We also commonly say that "b is premultiplied by a" or that "a is postmultiplied by b."

It is important to add "premultiplier" and "postmultiplier" to your vocabulary. It is also important to know that vector and matrix multiplication is not commutative. Later examples will clarify this point.

A ROW VECTOR TIMES A COLUMN VECTOR

Suppose the price list of the groceries purchased in the preceding problem is as follows.

bread	$.20 per loaf
meat	.90 per pound
potatoes	.05 each
lettuce	.40 each
ice cream	.80 per quart

The quantity of food purchased is expressed by the vector $(3, 6, 12, 3, 3)$. What is the total bill? Obviously the bill is calculated by multiplying each quantity by its associated price and adding up the products:

$$
\begin{array}{r}
\$ \;\;.60 \\
5.40 \\
.60 \\
1.20 \\
2.40 \\
\hline
\$10.20
\end{array}
$$

This is an example of row-by-column vector multiplication. This can be written with the price list as a row premultiplier and the quantity list as a column postmultiplier:

$$
(.20, .90, .05, .40, .80) \begin{pmatrix} 3 \\ 6 \\ 12 \\ 3 \\ 3 \end{pmatrix} = 10.20
$$

or, the vectors can be exchanged, for example,

$$
(3, 6, 12, 3, 3) \begin{pmatrix} .20 \\ .90 \\ .05 \\ .40 \\ .80 \end{pmatrix} = 10.20
$$

In either case, a row vector as a premultiplier multiplied by a column vector as a postmultiplier results in a single number (scalar). The result is the sum of the products of the corresponding elements in the two vectors. The two vectors *must* be of the same dimension. This multiplication can be written in general form. Let

$$u = (u_1, u_2, \ldots, u_n) \quad \text{and} \quad v = (v_1, v_2, \ldots, v_n)$$

Then

$$uv' = vu' = \sum_{i=1}^{n} u_i v_i$$

The use of the vector symbolization is often used in statistics to indicate sums of products, that is, one can write uv' or vu' instead of $\sum_{i=1}^{n} u_i v_i$, thus saving time by reducing the number of symbols needed.

Here is another very important statistical example. Let x be a column matrix such that $x' = (2, 3, 1, 4, 6)$. What is $x'x$?

$$x'x = (2, 3, 1, 4, 6) \begin{pmatrix} 2 \\ 3 \\ 1 \\ 4 \\ 6 \end{pmatrix}$$

$$= 2 \cdot 2 + 3 \cdot 3 + 1 \cdot 1 + 4 \cdot 4 + 6 \cdot 6$$

$$= 2^2 + 3^2 + 1^2 + 4^2 + 6^2 = 66$$

The form "$x'x$," where x is a column matrix, is the *sum of squares* of the elements of x.

PROBLEMS

Let $a = (1, 2, 4)$, $b = (3, 1, 1, 2)$, and $c = (1, 2, 4, 1)$, $d = \begin{pmatrix} 2 \\ -1 \\ 1 \end{pmatrix}$
What are the following?

12.8. ad

12.9. cd

12.10. bc'

12.11. $d'a'$

12.12. $d'd$

12.13. bb'

A COLUMN VECTOR TIMES A ROW VECTOR

The product of vectors when the premultiplier is a column vector and the postmultiplier is a row vector is a matrix. This topic is reserved as a special case in the next chapter. It was stated earlier that, in general, if a and b are vectors, then

$$a \times b \neq b \times a$$

Let a and b both be *column* vectors of the same order. Then we have the following relationships.

1. ab is undefined.   (column-by-column)
2. ba is undefined.   (column-by-column)
3. $a'b'$ is undefined.   (row-by-row)
4. $b'a'$ is undefined.   (row-by-row)
5. $a'b$ is a scalar.   (row-by-column)
6. ab' is a matrix.   (column-by-row)

A SPECIAL VECTOR PRODUCT

A vector of special interest is one that consists entirely of ones. A vector of ones is called a "unit vector." Here are some examples:

$$(1, 1, 1), \begin{pmatrix} 1 \\ 1 \\ 1 \\ 1 \end{pmatrix}, (1, 1, 1, 1, 1), \begin{pmatrix} 1 \\ 1 \end{pmatrix}$$

The unit vector is generally designated by the numeral "1" if it is a unit column vector. Unit row vectors are usually designated by "1'."

A column vector premultiplied by a unit row vector or a row vector post-multiplied by a column unit vector are ways of saying, "Add up the elements in the given vector." Thus, if $x = (1, 2, 4, 6)$ and we wish to write in vector notation the sum $\sum x_i$, we can write either

$$x1 \qquad \text{or} \qquad 1'x'$$

We have $x1 = 1'x' = \sum x_i = 13$.

Problems for Review

1. Suppose 10 persons are given two examinations—a verbal skills test and a mathematics skills test. Let scores be called "v_i" and "m_i" for $i = 1, \ldots, 10$; and let the score vectors be called "v" and "m."

Write in vector notation the following relationships. Assume v and m are column vectors.

a. The vector of total scores t, the elements of which are $t_i = m_i + v_i$.
b. The sum of verbal scores.
c. $\sum v_i m_i$
d. $\sum m_i^2$
e. $\sum t_i$

2. Suppose, in the above problem, the ten data vectors obtained are as follows:

[the pairs are ordered (v_i, m_i)] (10, 11)   (15, 10)

(6, 8)   (7, 9)

(9, 12)   (8, 4)

(8, 14)   (4, 9)

(6, 12)   (12, 6)

Put the data into a 10 × 2 matrix and carry out the following operations.

  a. t'
  b. $t'1$
  c. $v'm$
  d. $v' + 3m'$
  e. $v'v$
  f. $m'm$
  g. $1'm$

Matrix Algebra

Introduction

A MATRIX is a set of numbers arranged in a rectangular pattern. Essentially it can be several column vectors placed side by side or several row vectors placed one above the other. Since matrices can be thought of as sets of vectors and vectors can be considered special matrices, there should be many ways in which the concepts of the previous chapter can be extended to matrix algebra.

This is, in fact, the case. This chapter develops extensions of vector operations already discussed. Some new topics, however, are also brought up that are not generalizations of vector operations.

Matrix Addition and Subtraction

Matrix addition and subtraction is exactly the same as vector addition and subtraction and has the same limitations and properties.

Let A, B, and C be matrices defined as follows.

$$A = \begin{pmatrix} 2 & 1 \\ -3 & 4 \\ 1 & 2 \end{pmatrix} \quad B = \begin{pmatrix} 3 & -4 \\ 1 & 1 \\ -2 & 5 \end{pmatrix} \quad C = \begin{pmatrix} 0 & 3 \\ 1 & -1 \\ -2 & 4 \end{pmatrix}$$

We can now define the matrices $A + B$, $B + C$, and $A - C$. These are presented below.

$$A + B = \begin{pmatrix} (2 + 3) & (1 - 4) \\ (-3 + 1) & (4 + 1) \\ (1 - 2) & (2 + 5) \end{pmatrix} = \begin{pmatrix} 5 & -3 \\ -2 & 5 \\ -1 & 7 \end{pmatrix}$$

$$B + C = \begin{pmatrix} 3 & -1 \\ 2 & 0 \\ -4 & 9 \end{pmatrix}$$

$$A - C = \begin{pmatrix} 2 & -2 \\ -4 & 5 \\ 3 & -2 \end{pmatrix}$$

The additions and subtractions are done element by element just as was done in vector addition and subtraction.

Since corresponding elements in the two matrices are added, addition and subtraction are defined *only* for matrices of the same order.

PROBLEMS

Use the matrices A, B, and C as defined above.

13.1. Demonstrate the commutative principle by showing that

$$A + B = B + A \quad \text{and} \quad A + C = C + A$$

13.2. Demonstrate the associative principle by showing that

$$(A + B) + C = A + (B + C)$$

Matrix Multiplication

MULTIPLYING A MATRIX BY A SCALAR

A matrix multiplied by a scalar yields a new matrix, each element of which is the scalar times the corresponding elements of the original matrix. This is exactly analogous to scalar-times-vector products. Here are some examples.

$$A = \begin{pmatrix} 2 & 4 & 3 \\ 6 & 1 & 2 \end{pmatrix} ; \qquad\qquad 4A = \begin{pmatrix} 8 & 16 & 12 \\ 24 & 4 & 8 \end{pmatrix}$$

$$(\tfrac{1}{2})A = \begin{pmatrix} 1 & 2 & \tfrac{3}{2} \\ 3 & \tfrac{1}{2} & 1 \end{pmatrix}$$

$$B = \begin{pmatrix} b_{11} & b_{12} & b_{13} & b_{14} \\ b_{21} & b_{22} & b_{23} & b_{24} \end{pmatrix} ; \qquad kB = \begin{pmatrix} kb_{11} & kb_{12} & kb_{13} & kb_{14} \\ kb_{21} & kb_{22} & kb_{23} & kb_{24} \end{pmatrix}$$

PROBLEM   13.3

Let $U = \begin{pmatrix} 1 & 4 & 3 \\ 2 & 1 & 6 \\ 1 & 4 & 2 \\ 3 & 1 & 2 \end{pmatrix}$

Find $3U$ and $.6U$.

PRODUCTS OF VECTORS AND MATRICES

A matrix can be *premultiplied* by a row vector only if the order of the vector equals the number of rows in the matrix. The number of columns in the matrix is irrelevant. A matrix *cannot* be premultiplied by a *column* vector (except in the degenerate case that the matrix is really a row vector). The resulting product of a row vector times a matrix is a *row vector*, the elements of which are vector-vector products like the ones discussed in the previous chapter. The order of the resulting vector equals the number of columns in the original matrix.

Let's do some examples. Let $u = (1, 3, 2)$ and $X = \begin{pmatrix} 1 & 3 \\ 4 & 1 \\ 6 & 2 \end{pmatrix}$. Then the

two elements of uX are calculated by

$$1 \times 1 + 3 \times 4 + 2 \times 6 = 25$$

and

$$1 \times 3 + 3 \times 1 + 2 \times 2 = 10$$

yielding

$$uX = (25, 10)$$

Notice that there is one element in uX for each column of X and that u

has as many elements as X has rows. Also, if each *column* of X is considered a column vector—say $X = (x_1, x_2)$, where the x's are column vectors—then the elements of uX are each the vector products ux_1 and ux_2.

Here is another example. Let $x = (1, 1, 2, 2)$ and $Y = \begin{pmatrix} 3 & 4 \\ 1 & 3 \\ 2 & 1 \\ -1 & 1 \end{pmatrix}$

$$x Y = (1, 1, 2, 2) \begin{pmatrix} 3 & 4 \\ 1 & 3 \\ 2 & 1 \\ -1 & -1 \end{pmatrix}$$

$$= ((3 + 1 + 4 - 2), (4 + 3 + 2 - 2)) = (6, 7)$$

Note that x has as many elements (4) as Y has rows and that xY has as many elements (2) as Y has columns.

PROBLEMS

Let A, B, C, and D be vectors and matrices as follows.

$$A = \begin{pmatrix} 1 & 4 \\ 2 & 2 \\ 3 & 1 \\ 4 & 2 \end{pmatrix} \qquad B = \begin{pmatrix} 1 & 2 & 1 & 0 \\ 4 & 1 & 3 & 1 \\ 2 & 1 & 1 & 0 \\ 1 & 3 & 1 & 0 \end{pmatrix}$$

$$C = \begin{pmatrix} 1 & 3 \\ 2 & 6 \\ 2 & 4 \end{pmatrix} \qquad D = (4, 0, 1, 2)$$

Calculate each of the indicated products if it is defined.

13.4. DA
13.5. DB
13.6. DC

A matrix can be *postmultiplied* by a column vector only if the number of columns in the matrix equals the order of the column vector. The number of rows in the matrix is irrelevant. The resulting product is a *column vector*, the elements of which are vector-vector products as before. The order of the product vector equals the number of rows in the matrix. Here is an example.

$$\begin{pmatrix} 1 & 2 & 3 & 1 \\ 2 & 3 & 1 & 4 \\ 1 & 1 & 2 & 3 \end{pmatrix} \begin{pmatrix} 6 \\ 1 \\ 2 \\ 4 \end{pmatrix} = \begin{pmatrix} 1 \times 6 + 2 \times 1 + 3 \times 2 + 1 \times 4 \\ 2 \times 6 + 3 \times 1 + 1 \times 2 + 4 \times 4 \\ 1 \times 6 + 1 \times 1 + 2 \times 2 + 3 \times 4 \end{pmatrix} = \begin{pmatrix} 18 \\ 33 \\ 23 \end{pmatrix}$$

Notice that the product vector is a column vector of 3 elements and the original matrix has 3 rows. Notice that the postmultiplier has 4 elements and the matrix has 4 columns. Also, be sure you can see how the product vector was obtained by a series of vector-vector multiplications.

PROBLEMS

The last section defined A, B, C, and D. Let $E = \begin{pmatrix} 2 & 3 & 1 & 0 \\ 0 & 2 & 4 & 1 \end{pmatrix}$

Find each of these products, if they are defined.

13.7. AD'
13.8. BD'
13.9. CD'
13.10. ED'

PRODUCTS OF UNIT VECTORS AND MATRICES

The unit vectors 1 and 1′ can be used to indicate the sums of the elements in matrix rows and columns. Let A be

$$\begin{pmatrix} 2 & 1 & 4 & 6 \\ 1 & 3 & 1 & 2 \\ 2 & 1 & 4 & 1 \end{pmatrix}$$

Then $A1$ gives the column vector of row sums. We have

$$\begin{pmatrix} 2 & 1 & 4 & 6 \\ 1 & 3 & 1 & 2 \\ 2 & 1 & 4 & 1 \end{pmatrix} \begin{pmatrix} 1 \\ 1 \\ 1 \\ 1 \end{pmatrix} = \begin{pmatrix} 13 \\ 7 \\ 8 \end{pmatrix}$$

Also $1'A$ gives the row vector of column sums. We have

$$(1, 1, 1) \begin{pmatrix} 2 & 1 & 4 & 6 \\ 1 & 3 & 1 & 2 \\ 2 & 1 & 4 & 1 \end{pmatrix} = (5, 5, 9, 9)$$

The dimensionality of the unit vectors are determined by the matrix order.

PROBLEM 13.11

Let B be a $n \times m$ matrix of elements b_{ij}. Write in summation notation the products $B1$ and $1'B$.

THE PRODUCT OF TWO MATRICES

The foundation has been laid to consider the multiplication of two matrices. This will be done as several vector-vector multiplications. Any two matrices can be multiplied as long as the number of columns in the premultiplier equals the number of rows in the postmultiplier. If this condition does not hold, then the multiplication is not defined.

Let's take an example. Let

$$A = \begin{pmatrix} 1 & 2 & 3 \\ 4 & 5 & 6 \end{pmatrix} \quad \text{and} \quad B = \begin{pmatrix} 1 & 3 & 1 & 6 \\ 2 & 3 & 0 & 1 \\ 1 & 4 & 2 & 1 \end{pmatrix}$$

The product AB is defined but the product BA is not. Be sure you can determine this from the orders of B and A. AB is a (2×3) matrix times a (3×4) matrix—the number of columns in the premultiplier and the number of rows in the postmultiplier are equal (3).

Let's look at row one of A. Let's call this row vector a_1, where $a_1 = (1, 2, 3)$. We can multiply a_1 by B:

$$a_1 B = (8, 21, 7, 11)$$

We can also define the second row of A as a_2 where $a_2 = (4, 5, 6)$. The product $a_2 B$ can be calculated:

$$a_2 B = (20, 51, 16, 35)$$

The products $a_1 B$ and $a_2 B$ can be written together in matrix form as

$$AB = \begin{pmatrix} 8 & 21 & 7 & 11 \\ 20 & 51 & 16 & 35 \end{pmatrix}$$

So we see that the postmultiplication of a (2×3) matrix by a (3×4) matrix is precisely equivalent to doing 8 (i.e., 2×4) vector products and writing the results in the form of a (2×4) matrix.

Let's do some more examples. Let

$$C = \begin{pmatrix} 1 & 2 \\ 3 & 4 \end{pmatrix} \quad \text{and} \quad D = \begin{pmatrix} 3 & 1 & 2 \\ 2 & 4 & 1 \end{pmatrix}$$

$$CD = \begin{pmatrix} (1 \times 3 + 2 \times 2) & (1 \times 1 + 2 \times 4) & (1 \times 2 + 2 \times 1) \\ (3 \times 3 + 4 \times 2) & (3 \times 1 + 4 \times 4) & (3 \times 2 + 4 \times 1) \end{pmatrix}$$

$$= \begin{pmatrix} 7 & 9 & 4 \\ 17 & 19 & 10 \end{pmatrix}$$

Here we have a (2×2) matrix postmultiplied by a (2×3) matrix. It is a defined multiplication and is carried out as six vector-vector products, which are written in a (2×3) product matrix.

Let's construct a general summation rule to define matrix multiplication. Let X be a $(n \times k)$ matrix of elements x_{ij} and let Y be a $(k \times m)$ matrix of elements y_{jl}. Then the product XY is a $(n \times m)$ matrix of elements

$$\sum_{j=1}^{k} x_{ij} y_{jl}$$

Here is an example. Let X be (2×3) and Y be (3×2). Then XY is (2×2). We have

$$\begin{pmatrix} x_{11} & x_{12} & x_{13} \\ x_{21} & x_{22} & x_{23} \end{pmatrix} \begin{pmatrix} y_{11} & y_{12} \\ y_{21} & y_{22} \\ y_{31} & y_{32} \end{pmatrix} = \begin{pmatrix} \sum x_{1j} y_{j1} & \sum x_{1j} y_{j2} \\ \sum x_{2j} y_{j1} & \sum x_{2j} y_{j2} \end{pmatrix}$$

where all sums range from $j = 1$ to 3. You might want to check this out by using your own examples.

PROBLEMS

13.12. Suppose we have an assortment of matrices whose dimensions are as follows:

A is (4×5) D is (2×4)

B is (2×4) E is (5×2)

C is (4×2) F is (2×6)

Moreover, let G be a row vector of order 4 and H be a column vector of order 5. Indicate which of the following products are defined, and if so, what the orders of the products will be.

a. *AE*   g. *EF*
b. *BC*   h. *DF*
c. *BD*   i. *AE*
d. *CD*   j. *AEB*
e. *GA*   k. *BDC*
f. *AH*   l. *DAH*

Hint. It sometimes helps to use box diagrams for these determinations. For example, *CF* can be diagrammed as

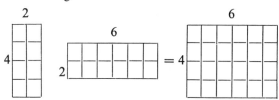

CF is seen to be defined and of order (4 × 6).

13.13. Matrices *P*, *Q*, *R*, and *S* are defined as follows

$$P = \begin{pmatrix} 1 & 2 & 1 \\ 3 & 1 & 2 \end{pmatrix} \qquad Q = \begin{pmatrix} 2 & 1 & 4 \\ 3 & 6 & 2 \\ 1 & 4 & 1 \end{pmatrix}$$

$$R = \begin{pmatrix} -1 & 3 & -4 \\ 2 & 1 & 1 \end{pmatrix} \qquad S = \begin{pmatrix} 3 & -2 \\ 1 & 4 \\ -1 & 2 \end{pmatrix}$$

Do the indicated matrix multiplications.

a. *PQ*   e. *SR*
b. *RQ*   f. *QS*
c. *PS*   g. *PQS*
d. *RS*   h. *RSP*

Matrix Transposition

INTRODUCTION

Vectors were transposed by changing rows into columns and columns into rows. Matrices are transposed in the same way. A transpose of a matrix is indicated by a prime, just as we used a prime to indicate transposed vectors. Let

$$A = \begin{pmatrix} 1 & 2 & 3 \\ 4 & 5 & 6 \end{pmatrix}$$

then

$$A' = \begin{pmatrix} 1 & 4 \\ 2 & 5 \\ 3 & 6 \end{pmatrix}$$

The first row of A is the first column of A'. The second row of A is the second column of A'. Similarly, the first row of A' is the first column of A. The second row of A' is the second column of A, and the third row of A' is the third column of A.

If we designate the rows of A by a_1 and a_2, where a_1 and a_2 are row vectors, then

$$A = \begin{pmatrix} a_1 \\ a_2 \end{pmatrix}$$

and

$$A' = (a_1', a_2')$$

In general, if X is a matrix of elements x_{ij}, then the $(i, j)^{th}$ element of X' is x_{ji} (the subscripts are reversed).

PROBLEM 13.14

Let

$$A = \begin{pmatrix} 1 & 2 & 3 & 4 \\ 3 & 1 & 2 & 1 \end{pmatrix} \qquad B = \begin{pmatrix} 2 & 1 & 3 & 4 \\ 1 & 2 & 1 & 3 \\ 4 & 1 & 2 & 5 \end{pmatrix}$$

Find A' and B'.

THE MATRIX PRODUCT $X'X$

A matrix form $X'X$ is of special interest since it corresponds roughly to the squaring operation. If X is a row vector, then $X'X$, as was seen earlier, is a sum of squares of the elements of X. However, if X is a matrix, then $X'X$ shows sums of squares and sums of cross products of the *columns* of X. Matrices of the form $X'X$ are of special interest to statisticians, as you will learn later. Such matrices are *always* symmetric. This fact will help you be sure that the calculations are correct, and can shorten the work involved. Let's do an example. Let

$$A = \begin{pmatrix} 1 & 3 \\ 2 & 4 \\ 0 & 3 \\ 2 & 1 \\ 3 & 2 \end{pmatrix}$$

$$A'A = \begin{pmatrix} (1^2 + 2^2 + 0^2 + 2^2 + 3^2) & (1\cdot3 + 2\cdot4 + 0\cdot3 + 2\cdot1 + 3\cdot2) \\ (1\cdot3 + 2\cdot4 + 0\cdot3 + 2\cdot1 + 3\cdot2) & (3^2 + 4^2 + 3^2 + 1^2 + 2^2) \end{pmatrix}$$
$$= \begin{pmatrix} 18 & 19 \\ 19 & 39 \end{pmatrix}$$

Do you see how the elements of $A'A$ are constructed by summing the squares and cross products of the *columns* of A? Let's do another example. Let

$$B = \begin{pmatrix} 3 & 4 & 1 \\ 2 & 0 & 1 \\ 3 & 1 & 2 \end{pmatrix}$$

$$B'B = \begin{pmatrix} 3 & 2 & 3 \\ 4 & 0 & 1 \\ 1 & 1 & 2 \end{pmatrix} \begin{pmatrix} 3 & 4 & 1 \\ 2 & 0 & 1 \\ 3 & 1 & 2 \end{pmatrix} = \begin{pmatrix} 22 & 15 & 11 \\ 15 & 17 & 6 \\ 11 & 6 & 6 \end{pmatrix}$$

Note that $B'B$ is symmetric. Also be sure you see that the elements of $B'B$ are sums of squares (on the diagonal) and sums of cross products (off the diagonal).

PROBLEMS

13.15. If C is of the order (7×4), what is the order of $C'C$? What is the order of CC'? Is CC' also symmetric?

13.16. If X is a matrix of elements x_{ij} and is of order N by K, write in summational form the third element of the second row of $X'X$.

Matrix Inversion

Matrix inversion is the matrix analog to reciprocals of scalars. It is used as a sort of matrix division device. Notice that "$65 \div 5$" means the same thing as "$65 \cdot \frac{1}{5}$" and also keep in mind that $5 \cdot \frac{1}{5} = 1$. Recall also the definition of the identity matrix. The identity matrix is the matrix analog of the number one. It is a diagonal matrix of zeros and ones and is always denoted by "I." It has this general pattern.

$$I = \begin{pmatrix} 1 & 0 & 0 \\ 0 & 1 & 0 \\ 0 & 0 & 1 \end{pmatrix}$$

The *inverse* of a matrix is designated by a negative one exponent. Thus, "A^{-1}" means "the inverse of A." Now, let's define "A^{-1}." For a given matrix named "A," A^{-1} is the matrix that makes these equations true:

$$AA^{-1} = I \quad \text{and} \quad A^{-1}A = I$$

Every matrix has what is called a "generalized inverse," however, generalized inverses are not needed until one gets into advanced, special statistical analyses. Usually we need only real inverses as defined.

Specifically, let's accept categorically that "inverse" is a concept reserved for *square* matrices. The inverse will also be a square matrix. Moreover, *not* all square matrices have inverses (or, as is sometimes said, "not all square matrices can be inverted").

We will not deal at all in this text on distinguishing invertable from non-invertable matrices, nor will we deal much with determining inverses. It should be pointed out that finding an inverse is, in general, a matter of solving several sets of systems of simultaneous equations. Let's see how this is.

Let

$$A = \begin{pmatrix} 4 & 1 \\ 3 & 2 \end{pmatrix}$$

We want to find a matrix A^{-1} such that $A^{-1}A = I$. We know that A^{-1} must be square and of order (4×4). Letting the elements of A^{-1} be designated by a, b, c, and d, we can write our goal as finding a, b, c, and d such that

$$\begin{pmatrix} 4 & 1 \\ 3 & 2 \end{pmatrix}\begin{pmatrix} a & b \\ c & d \end{pmatrix} = \begin{pmatrix} 1 & 0 \\ 0 & 1 \end{pmatrix}$$

is a true statement. We can decompose this matrix equation into two sets of two simultaneous scalar equations in two unknowns each. These are

$$4a + c = 1 \qquad 4b + d = 0$$
$$3a + 2c = 0 \qquad 3b + 2d = 1$$

Be sure you see how to obtain these equations from matrix multiplication.

The first two equations can be solved simultaneously to yield $a = .4$, and $c = -.6$. The second set yields $b = -.2$ and $d = .8$. Let's check and see if we have an inverse.

$$\begin{pmatrix} 4 & 1 \\ 3 & 2 \end{pmatrix} \begin{pmatrix} .4 & -.2 \\ -.6 & .8 \end{pmatrix} = \begin{pmatrix} 1 & 0 \\ 0 & 1 \end{pmatrix}$$

Our computations check out, so we have in fact found A^{-1} (or, we have "inverted A"). You might also wish to see if $A^{-1}A = I$.

Now, consider

$$B = \begin{pmatrix} 4 & 2 & 2 \\ 3 & 1 & 0 \\ 2 & 0 & 4 \end{pmatrix}$$

We can find B^{-1} similarly. Let

$$B^{-1} = \begin{pmatrix} a_1 & b_1 & c_1 \\ a_2 & b_2 & c_2 \\ a_3 & b_3 & c_3 \end{pmatrix}$$

We can construct the following three sets of three simultaneous equations in three unknowns:

$$4a_1 + 2a_2 + 2a_3 = 1$$
$$3a_1 + a_2 \qquad\quad = 0$$
$$2a_1 + \qquad\quad 4a_3 = 0$$

$$4b_1 + 2b_2 + 2b_3 = 0$$
$$3b_1 + b_2 \qquad\quad = 1$$
$$2b_1 + \qquad\quad 4b_3 = 0$$

$$4c_1 + 2c_2 + 2c_3 = 0$$
$$3c_1 + c_2 \qquad\quad = 0$$
$$2c_1 + \qquad\quad 4c_3 = 1$$

Each of these sets of equations will give three of the elements of B^{-1}. The first set yields $a_1 = -\frac{2}{6}$, $a_2 = 1$, and $a_3 = \frac{1}{6}$. Find the other six elements and check your answers by the formula

$$BB^{-1} = I$$

The procedure can be generalized to the inversion of larger matrices, but these procedures get tedious and mistakes are easy. There are numerous methods to speed up and systemize the work, and inversion is done fairly easily on computers.

Equation Systems in Matrix Form

Suppose we have the task of solving the equations

$$4x + 2y - 8 = 0$$
$$3x + y + 2 = 0$$

Let's rewrite these in a standard form

$$4x + 2y = 8$$
$$3x + y = -2$$

These equations can be written in matrix form as

$$\begin{pmatrix} 4 & 2 \\ 3 & 1 \end{pmatrix}\begin{pmatrix} x \\ y \end{pmatrix} = \begin{pmatrix} 8 \\ -2 \end{pmatrix}$$

The solution to x and y can be obtained by premultiplying both sides of the equation by $\begin{pmatrix} 4 & 2 \\ 3 & 1 \end{pmatrix}^{-1}$ We obtain

$$\begin{pmatrix} 4 & 2 \\ 3 & 1 \end{pmatrix}^{-1}\begin{pmatrix} 4 & 2 \\ 3 & 1 \end{pmatrix}\begin{pmatrix} x \\ y \end{pmatrix} = \begin{pmatrix} 4 & 2 \\ 3 & 1 \end{pmatrix}^{-1}\begin{pmatrix} 8 \\ -2 \end{pmatrix}$$

Since

$$\begin{pmatrix} 4 & 2 \\ 3 & 1 \end{pmatrix}^{-1}\begin{pmatrix} 4 & 2 \\ 3 & 1 \end{pmatrix} = \begin{pmatrix} 1 & 0 \\ 0 & 1 \end{pmatrix}$$

and

$$\begin{pmatrix} 1 & 0 \\ 0 & 1 \end{pmatrix}\begin{pmatrix} x \\ y \end{pmatrix} = \begin{pmatrix} x \\ y \end{pmatrix}$$

we obtain

$$\begin{pmatrix} x \\ y \end{pmatrix} = \begin{pmatrix} 4 & 2 \\ 3 & 1 \end{pmatrix}^{-1}\begin{pmatrix} 8 \\ 2 \end{pmatrix}$$

So the solution depends on finding the inverse. The inverse is

$$\begin{pmatrix} -\tfrac{1}{2} & 1 \\ \tfrac{3}{2} & -2 \end{pmatrix}$$

and the solution is

$$\begin{pmatrix} x \\ y \end{pmatrix} = \begin{pmatrix} -\tfrac{1}{2} & 1 \\ \tfrac{3}{2} & -2 \end{pmatrix}\begin{pmatrix} 8 \\ -2 \end{pmatrix} = \begin{pmatrix} -6 \\ 16 \end{pmatrix}$$

or $x = -6$ and $y = 16$.

The solution to two simultaneous equations is simplified by observing that the inverse of a two-by-two matrix is quite simple. Let $X = \begin{pmatrix} a & b \\ c & d \end{pmatrix}$, then

$$X^{-1} = \frac{1}{ad - bc}\begin{pmatrix} d & -b \\ -c & a \end{pmatrix}$$

Try it. Find $\begin{pmatrix} 2 & 1 \\ 4 & 3 \end{pmatrix}^{-1}$. The answer is

$$\frac{1}{2 \cdot 3 - 1 \cdot 4}\begin{pmatrix} 3 & -1 \\ -4 & 2 \end{pmatrix} = \begin{pmatrix} \frac{3}{2} & -\frac{1}{2} \\ -2 & 1 \end{pmatrix}$$

PROBLEMS

Write the following equation systems in matrix form. Show the equations that give the solutions.

13.17. $4a + 2b + c = 10$
$3a + b + 2c = 1$
$a + 2b + 3c = 5$

13.18. $2x_1 + 10x_2 - 2x_3 = 8$
$- 3x_2 + 4x_3 = 4$
$5x_1 - 2x_2 + 6x_3 = 8$

Problems for Review

Here are some matrices and vectors that will be used in the problems below.

$$A = \begin{pmatrix} 2 & 1 & 2 & 3 \\ 3 & 1 & 2 & 1 \end{pmatrix} \quad B = \begin{pmatrix} 1 & 4 \\ 0 & -2 \\ 1 & 1 \\ 4 & 0 \end{pmatrix} \quad C = \begin{pmatrix} 1 & 2 \\ 3 & -2 \end{pmatrix}$$

$$D = \begin{pmatrix} 1 \\ 2 \\ 4 \\ 1 \end{pmatrix} \quad E = \begin{pmatrix} -2 \\ 1 \end{pmatrix}$$

1. Determine the following matrices.

a. $A - 2B'$
b. BC
c. $D'B$
d. $1'B1$
e. C^{-1}
f. $B'B$

g. $(B'B)^{-1}$
h. $(D'D)^{-1}$
i. DD'
j. EE'
k. $AB + 2C$
l. $C - (1'C1)I$

2. Suppose there exists a state with 6 counties. Each county has 4 state taxes to collect and taxes are collected twice a year. Let's construct two 6 × 4 matrices to reflect the tax records (in thousands of dollars).

| | T_s | | | | | T_w | | | |
| | Summer Tax Source | | | | | Winter Tax Source | | | |
County	Income	Sales	Property	Gas		Income	Sales	Property	Gas
1	5	5	1	1		6	6	1	2
2	10	6	6	2		8	5	5	3
3	3	4	7	1		2	5	8	1
4	50	10	20	3		40	8	25	5
5	4	1	5	1		4	2	4	1
6	90	20	40	4		70	20	50	5

Construct the following matrices and vectors. Give formulas and numerical solutions.

a. The matrix showing, by county, the amount of each tax gathered in the entire year.

b. The sum of the summer taxes for the state by tax category.

c. The sum of the winter taxes for each county.

d. The total year's tax revenue for the state.

e. The matrix showing the sum of squares and cross products for each tax source for the summer tax data.

Part Four

SET ALGEBRA AND PROBABILITY

THIS part is about the mathematical language and concepts called "set theory." The ideas embodied here are important primarily as a tool for studying probability theory, which is the foundation of statistics. There are also other reasons why set concepts should be learned, and some of these will be pointed out as we progress.

The learning of statistical *understanding* is greatly facilitated by an understanding of probability basics. However, the learning of probability concepts is greatly handicapped if the learner has no grasp of basic set theory concepts. The use of the set language and set algebra simplifies probability concepts considerably.

Set concepts have not always been used in the teaching of applied statistics. However, it is becoming more common in modern approaches to the study of statistics because knowledge of probability concepts is so important in the understanding of statistical concepts. Probably many readers will need to learn only the terms—but these terms should be learned so thoroughly that they are completely understood if used orally by the statistics instructor. Some readers will need to study further—to learn the standard set symbols and operations. The reader who aspires to a research career, and who intends to study applied statistics seriously, will need a firm grounding in the set algebra fundamentals.

Some students will want to go into much greater detail than this text allows. These students will profit by studying other texts, such as McGinnis' *Mathematical Foundations for Social Analysis* (Chapters 1–4) and Kemeny, Snell, and Thompson's *Introduction to Finite Mathematics*.

Basic Set Theory Concepts

Introduction

THIS chapter deals primarily with set language and notation, the knowledge of which is of utmost importance. Every new term and symbol introduced in this chapter should be made part of your basic research vocabulary.

Basic Definitions

SETS

The idea of a set is so fundamental in mathematics that it can be used as a starting point for developing the entire field. Once you have an intuitive idea of what a set is, the set concept becomes a very powerful tool in understanding mathematics in general and probability and statistics in particular.

Since the idea of a set is so fundamental, it is difficult to define. A frequently used definition is, "a set is a collection of things." This is a tautology since one could just as easily say "a collection is a set of things." So the term "set" shall remain, at best, poorly defined.

Let's expand the definition of "set" and say:

A set is a well-defined collection of things.

The insertion of "well-defined" means that a *rule* exists whereby we can tell whether or not something belongs in the set.

Although this definition is still a tautology, an intuitive idea of sets can be developed by examples, which can be presented in two ways. We can *list* the entire set. For example:

The set of all vowels is a, e, i, o, and u.

Or, we can *describe* the set without listing it. For example:

The set is the set of all vowels.

In some cases, when the entire set cannot be listed, the rule or description will have to suffice. For example, the set of all living human beings cannot be listed, but we can tell by observing an object whether or not it belongs to the set of all living human beings.

ELEMENTS

The word *"element"* is the technical word for the members of a set. The letters "a" and "e" are two elements of the set of all vowels. You are an element of the set of all living human beings.

NOTATION AND SYMBOLS

There are many common ways to symbolize sets and their elements. The use of braces is one common method. Thus, we can write

{a, e, i, o, u}

which is read as "the set consisting of the letters a, e, i, o, and u," or we can write

{all vowels}

which is read as, "the set consisting of all of the vowels."

Let's consider a simple statistics example. Suppose you are interested in determining the average height of the local basketball team. The team members make up a set of persons (or the *sample*, in statistical language). The set is, for example,

{John, Tom, Bob, Jack, Henry}

The heights of the five team members also compose a set. This set (data, in statistical language) might be in inches:

{72, 70, 76, 72, 75}

This set might or might not be in the same order as the team member set. The average, 73, is determined from the set of data.

Keep in mind that these two examples are *different* sets. The elements of the first set are people, while the elements of the second set are numbers (heights). Of course, they are related sets, but they are different.

Now, more shorthand will be helpful. The phrase

"set of all members of the local basketball team"

or even

"John, Tom, Bob, Jack, Henry"

takes considerable effort to write out or say. It is convenient to let one symbol represent the entire set. We can say

"let $T = \{$John, Tom, Bob, Jack, Henry$\}$"

or

"let $T = \{$members of the local basketball team$\}$"

In addition we might say

"let $H = \{72, 70, 76, 72, 75\}$"

Capital letters are usually used to designate sets. Whenever possible it is helpful if the letter chosen has some relationship to the set membership. In our example, T was chosen for the team and H was chosen for the height set.

It is also helpful to have a symbol to show whether or not something is an element of a particular set. The symbol "$\in$" is used universally to designate set membership. For example, "John $\in T$" means "John is an element of T." We can also write

$$76 \in H$$

and

$$e \in \{\text{all vowels}\}$$

Moreover, it is helpful to have a symbol to designate nonmembership. The slash (/) is used universally to negate relationships. Therefore, a slash is used to negate the set membership relationship. The slash is superimposed to yield "$\notin$". This strange symbol means "not an element." We have

$$\text{Pete} \notin T$$
$$64 \notin H$$
$$s \notin \{\text{all vowels}\}$$

Small-case letters are frequently used to designate the elements of sets. The sentence

"$t \in T$"

is read

"t is an element of T"

The small-case t might represent a particular element, e.g., it could be Tom, or a general element, that is, any one of the team members.

A different letter can be used for each element, but this can get awkward if there are too many elements. A subscript notation is usually employed. We might say

$$H = \{h_1, h_2, h_3, h_4, h_5\}$$

when

$$h_1 = 72, h_2 = 70, h_3 = 76, h_4 = 72, h_5 = 75$$

SUBSETS

Frequently, it is valuable to treat some elements of a set as a distinct set-within-a-set. The basketball coach might want to consider the two boys whose heights are less than six feet as a separate set. Such a set-within-a-set is called a *subset*.

Suppose the two short boys are given the guard positions. Let G stand for the set consisting of the two guards. We say

"G is a subset of T"

The symbol for the expression ". . . is a subset of . . ." is "$\subseteq$." The sentence

"$G \subseteq T$"

is read

"G is a subset of T"

Notice that the subset is at the closed end of the symbol and the parent set is at the open end of the symbol. The symbol is quite similar to the inequality symbols ($\leq$ and $\geq$). The open end of the inequality sign designates the larger quantity, just as the open end of the subset symbol ($\subseteq$) designates the set that is usually the larger set. The underline portion of the subset symbol ($\subseteq$) is used to imply that a subset and its parent set could be identical. For example, let S be the set of all local basketball players under 78 inches tall. Then $S \subseteq T$ and $S = T$, that is, S is a subset of T, but all elements of T are in S. T and S are identical sets, although their defining properties are different.

Let's take another example. Let

$$U = \{1, 2, 3, 4\}$$

Some subsets of U are $\{1, 2\}$, $\{1, 3, 4\}$, $\{4, 2\}$, and $\{3\}$.

A special subset of U is U (just like our sets T and S, above). What are some other subsets of U? How many subsets of U are there?
[Answer—15 (plus one more to be discussed later—the null set). There is one subset having four elements, four subsets having three elements, six subsets having two elements, four subsets having one element, and one subset having no elements.]

A subset that is definitely *not identical* to its parent set is called a *proper subset* of the parent set. The statement "$D \subset R$" means "D is a *proper subset* of R." The symbol for a proper subset is the same as the symbol for a subset except that the underline is not used.

Examples. Here are some examples of subsets of sets and proper subsets of sets.

Subsets:

   a. $\{5, 6, 1\} \subseteq \{1, 7, 6, 5, 4\}$

   b. $\{1, 2, 3\} \subseteq \{3, 2, 1\}$

   c. $\{a, b, d\} \subseteq \{a, b, c, d, e\}$

   d. {all men} $\subseteq$ {all people}

   e. $\{2, 4, 6, 8\} \subseteq$ {all even digits}

Proper Subsets:

   Which of the above examples are proper subsets? [*Answer.* a, c, d, and e.]

PROBLEM 14.1

   The following problems consist of pairs of sets. The task is to insert a symbol between the two sets so that each problem becomes a *true sentence.* Many of the problems have *more than one answer*, so think carefully about each problem. Determine *the complete set* of possible answers.

   a. $\{1, 3, 5)$ $\{9, 7, 5, 3, 1\}$
   b. 5 $\{1, 2, 3, 4, 5\}$
   c. {all odd numbers} {all numbers}
   d. $\{6, 2\}$ {all odd numbers}
   e. 6   {all odd numbers}

NULL SETS

   A *null set* has no elements and is frequently called an *empty set.*
   Some example sets that are null sets are:

   {all purple cows}
   {numbers which are *both* larger than 10 *and* smaller than 5}
   {all males over 12 feet tall}

Another set that is probably a null set is

   {all completely true generalizations}

   The idea of a null set is a simple, but necessary, concept. A null set is, first of all, a *set.* It is *not* the number zero. *It is a set that has zero* (i.e., no) *elements.*

Consider a fairly practical example.

Let

$$U = \{\text{all students enrolled in Stat. 200}\}$$
$$P = \{\text{all elements of } U \text{ that pass}\}$$
$$F = \{\text{all elements of } U \text{ that fail}\}$$

We have

$$P \subseteq U$$
$$F \subseteq U$$

If all students pass the course, then $P = U$ and F is a null set. The fact that F has no elements does not keep us from defining F and discussing it as a meaningful set.

COMPLEMENT OF A SET

Consider a set that we will call B. Suppose A is a subset of B. The *complement* of set A is a set consisting of elements that *do not* belong to A, but *do* belong to B. As an example, let B be the set of all numbers. Let A be all numbers greater or equal to zero. We can write:

$$A = \{\text{all } x \text{ such that } x \geq 0\}$$

The complement of A is the set of all numbers less than zero. The symbol "$\sim$" is used above a set symbol to denote the complement of the set. So the complement of A is denoted $\tilde{A}$. $\tilde{A}$ can be written as

$$\tilde{A} = \{\text{all } x \text{ such that } x < 0\}$$

Note that A and $\tilde{A}$ *taken together* consist of *all numbers*, that is, set B.

The example from the previous section will also be helpful. In this case $P = \{\text{students who passed Stat. 200}\}$ and $F = \{\text{students who failed Stat. 200}\}$. Assuming that there were no dropouts or audits,

$$P = \tilde{F}$$
$$F = \tilde{P}$$

That is, P and $\tilde{P}$ taken together, *or* F and $\tilde{F}$ taken together, make up U, where U is the set of all students enrolled in Stat. 200.

THE UNIVERSAL SET

Frequently you will see the term *universal set*. A set and its complement taken together are often called a universal set. In the previous section, sets B and U could have been called *universal sets*.

Problems for Review

1. Translate the following symbolic sentences into good English sentences.

a. $5 \in A$
b. {all even numbers} $\subset$ {all numbers}
c. $\{5, 7, 10\} \nsubseteq \{5, 7, 20, 24\}$
d. $A = \{2, 3, 4, 5, 6, 7\}$
e. $\tilde{D} \subset R$
f. $7 \notin \tilde{U}$

2. Define the following terms and symbols.

a. Null set.
b. Complement.
c. Element.
d. $\notin$.
e. $\in$.
f. $\subseteq$.

3. Answer the following questions.

a. What is the implied *complement* of {all males}?
b. What are all of the *proper subsets* of $\{h, k, n\}$?
c. What is the *difference* between the null set and a set A defined as $A = \{0\}$?
d. Suppose a poll is taken to determine if persons plan to vote Republican, Democratic, or Socialist in the next presidential election. Suppose 45% say Republican and 55% say Democrat. Describe the results in set language—use the terms "complement" and "null set."
e. Suppose the following data are gathered in a survey. The numbers are frequency counts based on a sample of 100 persons.

	Males	Females	
College graduates	20	10	30
Noncollege graduates	25	45	70
	45	55	

(1) How many sets of people are implied by the data?
(2) How many elements are in the set of all males?

(3) How many elements are in the complement of the set of all college graduates?

(4) Code the data by assigning letters to represent the rows and columns of the table. (Only two letters are needed.)

(5) Suppose the letter A is the set of 20 persons in the first cell. What was the rule by which members were assigned to A?

(6) How many elements are in $\tilde{A}$?

Set Operations

Introduction

THE preceding chapter introduced some set language and notation. This chapter introduces elementary set algebra. The ideas of this chapter will be useful in the calculation of probabilities and the clarification of statistical sampling problems.

There are two general types of set algebra problems that must be distinguished. One type deals with set operations and set membership. The second type deals with the number of elements in sets and in set combinations. This chapter is concerned with the former problem; the latter is covered in Chapter 16.

Set operations problems such as those in this chapter, are characterized by the manipulation of *sets*. The answer is usually a *set*, not a number. Problems such as those in the succeeding chapter are characterized by manipulations of *numbers* of elements in sets. The answer to these problems is usually a *number*, not a set.

Set Operations

INTERSECTION

Consider any two sets A and B. The *intersection* of A and B is a *set* whose elements consist of any element that is in *both* A and B. The idea of the *intersection* of two sets is easily understood through examples.

1. The intersection of $\{1, 2, 4\}$ and $\{2, 4, 8\}$ is $\{2, 4\}$.
2. The intersection of $\{a, e, i, o, u, y\}$ and $\{all\ consonants\}$ is $\{y\}$.
3. The intersection of $\{all\ males\}$ and $\{all\ school\ teachers\}$ is $\{all\ male\ school\ teachers\}$.

Note that an element belongs in the intersection of two sets *if it is in both* sets. The key word is "*and.*" If $a \in A$ *and* if $a \in B$, then a is in the intersection of A and B.

The term "intersection" can be understood easily by considering its most common meaning—the place where two streets cross. Let A be a set of cars traveling on street A. Let B be the set of cars traveling on street B. If *any* cars are in *both* sets A and B, they must be in the *intersection* of streets A and B!

The symbol for intersection looks like an inverted capital *u*. It is "$\cap$." We can write

$$\text{"} A \cap B = C \text{"}$$

which is read

$$\text{"The intersection of } A \text{ and } B \text{ is } C.\text{"}$$

The three previous examples can be written:

1. $\{2, 4, 8\} \cap \{1, 2, 4\} = \{4, 2\}$
2. $\{y\} = \{\text{all consonants}\} \cap \{a, e, i, o, u, y\}$
3. $\{\text{all males}\} \cap \{\text{all school teachers}\} = \{\text{all male school teachers}\}$

Here are some more examples of intersections.

1. $\{1, 2, 3, 4, 5, 6\} \cap \{2, 4, 6, 8, 10\} = \{2, 4, 6\}$
2. $\{\text{all } x \text{ such that } x > 10\} \cap \{\text{all } x \text{ such that } x < 12\} = \{\text{all } x \text{ such that } 10 < x < 12\}$
3. $\{\text{names of football positions}\} \cap \{\text{names of basketball positions}\} = \{\text{guard, center}\}$

DISJOINT SETS

Many pairs of sets have *no* elements in common. The intersection of two sets that have *no* elements in common is the *null set*. That is, every pair of sets has an intersection set, but in some cases, the intersection set has *no* elements. For example, the intersection of

$$\{2, 4, 6\} \quad \text{and} \quad \{1, 3, 5, 7\}$$

is the null set. The null set is also the intersection of

$$\{\text{all males}\} \quad \text{and} \quad \{\text{all females}\}$$

Such pairs of sets are said to be *disjoint* sets. Two sets are *disjoint* if they have no elements in common or, equivalently, if their intersection is the *null set*.

UNION

Consider any two sets, *A* and *B*. The *union* of *A* and *B* is a *set* whose elements consist of any element that is in *either A or B* (or both). The key word is "*or*." An element belongs in the *union* of two sets if it is in either one set *or* the other. Let's look at some examples.

1. The union of {1, 2, 4} and {6, 8} is {1, 2, 4, 6, 8}.
2. The union of {1, 2, 4} and {2, 4, 6} is {1, 2, 4, 6}.
3. The union of {all x such that $x > 10$} and {all x such that $x > 20$} is {all x such that $x > 10$}.
4. If A = {all families whose salaries exceed $3000} and if B = {all families who own their own homes}

then the union of *A* and *B* includes *all* families who *either* own their own home *or* who earn more than $3,000. It *does* include the very poor who own their own homes and it *does* include the very rich that rent. It does *not* include the very poor that rent.

The symbol for union looks like a capital *u*. It is "∪." We can write

$$“A \cup B = C”$$

which is read

$$\text{“The union of } A \text{ and } B \text{ is } C.”$$

PROBLEM 15.1

Some more examples, using the union symbol (∪) are:

a. {1, 2, 3} ∪ {1, 3, 5} = {1, 2, 3, 5}.
b. {i, j, k} ∪ {k, l, m} = {m, l, k, i, j}.
c. {all even numbers} ∪ {all odd numbers} = {all counting numbers}.
d. {all x such that $x > 0$} ∪ {all x such that $x < 0$} = {all numbers *except* 0}.
e. {all x such that $x \geq 0$} ∪ {all x such that $x \leq 10$} = {all numbers}.

It is important to distinguish between *union* and *intersection*. In the above five examples, find the intersections of the two given sets.

Compare the concepts of union and intersection and be sure you can tell whether or not an element belongs in the union or in the intersection or in both. (Notice that elements in the intersection of two sets are always in the union of the two sets.)

Circle Diagrams

A very helpful technique for finding unions and intersections is the circle diagram. Circle diagrams are usually called *Venn diagrams* or *Euler diagrams,* and are used frequently in the application of set ideas to the study of formal logic.

We can represent two general sets *A* and *B* by two overlapping circles. Shading certain areas clarifies the set relationships.

Figure 15.1 shows sets *A* and *B* as overlapping circles. The total shaded area represents the union, $A \cup B$; while the doubly shaded area represents the intersection, $A \cap B$. Try some examples. Draw circle diagrams for the following pairs of sets.

1. $A = \{a, b, c, d\}$, $B = \{a, c, g, h\}$
2. $A = \{1, 2, 3, 4\}$, $B = \{2, 4, 6\}$

In your first figure, the letters "a" and "c" should be in the football-shaped intersection area. The second figure should show the numerals "2" and "4" in the intersection.

Now take a more practical example. Draw overlapping circles. Let one be the set of all sophomores. Let the other be the set of all sociology majors. Think about the various areas of your diagram. Describe the intersection and union. What students are in the intersection? (Answer—sophomore sociology majors.) What students are in the union? What students are *outside* the two circles? What are the *complements* of *all* of the sets (remember that the union and the intersection are also sets)?

Another practical example is given by the final problem in the previous chapter. Look at that example. You can think of this problem as a complex circle diagram where squares have been drawn instead of circles. How many

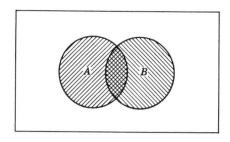

Figure 15.1   Circle diagrams.

persons are in the *intersection* of *C* (college graduates) and *M* (males)? (Answer—20) How many students are in the *union* of *C* and *M*? Careful! It is *not* 100. (Answer—55)

Perhaps these last two examples can give you a rough idea about how set language can help the statistician understand the composition of his sample (a statistical sample is the *set* of "things" that a statistician is studying.)

Cartesian Products of Sets

One set operation is quite useful in describing relations, correspondence, and functions from the set theory point of view. This operation is called the "Cartesian product of sets" and is indicated by the standard multiplication operator "×."

Let *A* and *B* be two sets. Let's pair each element of *A* with each element of *B* and thereby obtain all possible pairs of elements between the two sets. This *set* of all possible element *pairs* is the Cartesian product, $A \times B$.

Here is an example. If

$$A = \{1, 2\} \quad \text{and} \quad B = \{3, 5, 7\}$$

then

$$A \times B = \{(1, 3), (1, 5), (1, 7), (2, 3), (2, 5), (2, 7)\}$$

See how $A \times B$ contains all possible pairings of the elements of *A* with the elements of *B*.

Here are some more examples.

$$A = \{1, 2\}$$
$$D = \{a, b, c, d\}$$
$$E = \{\text{male, female}\}$$
$$A \times D = \{(1, a), (1, b), (1, c), (1, d), (2, a), (2, b), (2, c), (2, d)\}$$
$$A \times E = \{(1, \text{male}), (1, \text{female}), (2, \text{male}), (2, \text{female})\}$$

PROBLEM 15.2

What is $D \times E$?

Notice the number of elements (pairs) in the Cartesian product. The number of elements in $A \times D$ is 8; in $A \times E$, it is 4; and in $D \times E$, the number of pairs is 8. In each case, the number of elements in the Cartesian product is the product of the number of elements in each set.

"Relation" can be defined as any subset of a Cartesian product. For example, in the problem $A \times E$, the first and last element form a subset that define a relation between sex and the digits "1" and "2." The relation can be stated as the set

$$\{(1, \text{male}), (2, \text{female})\}$$

Also, the subset $\{(2, \text{male}), (1, \text{female})\}$ defines a similar relation. Both of these subsets can be used to code data on the sex of experimental subjects. Other subsets can be chosen from $A \times E$, none of which would have the utility of the two just discussed. One of the jobs of the statistician could be described as studying Cartesian products and trying to decide (or discover) which subsets are useful relations.

Problems for Review

1. Let A = all students whose CEEB V scores[1] exceed 500

B = all students CEEB Q scores[1] exceed 500

$C = A \cup B$

$D = A \cap B$

Here is a set of eight statements about sets A, B, C, and D. Find from among the eight statements the subset of true statements.

a. A student whose V score is 400 *might* be in C.
b. A student whose V score is 400 *might* be in D.
c. *All* students in C have a V + Q total of *at least* 1000.
d. *All* students in D have a V + Q total of *at least* 1000.
e. All students in $\tilde{A}$ are in $\tilde{C}$.
f. All students in $\tilde{A}$ are in $\tilde{D}$.
g. A student who has V = 400 and Q = 300 is in $\tilde{C}$.
h. A student who has V = 400 and Q = 600 is in $\tilde{D}$.

2. The following list contains *true* sentences about set relations. Study the list and convince yourself of the validity of the statements. You can use circle diagrams to help in several cases.

a. $A \cup (B \cup C) = (A \cup B) \cup C$
b. $A \cap (B \cap C) = (A \cap B) \cap C$
c. $A \cap (B \cup C) = (A \cap B) \cup (A \cap C)$
d. $A \cup (B \cap C) = (A \cup B) \cap (A \cup C)$

[1] Verbal (V) and Quantitative (Q) scores on the aptitude portion of the College Entrance Examination Board (CEEB) exams.

3. Consider Figure 15.2.

The figure consists of three overlapping circles representing three sets. The letters in the circles are not intended to be elements of the sets, but are merely *names* for the seven mutually disjoint *subsets* in the diagram. Suppose A, B, and C are as follows:

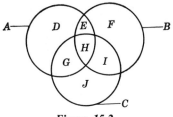

Figure 15.2

$$A = \{\text{all New Yorkers}\}$$
$$B = \{\text{all middle class Americans}\}$$
$$C = \{\text{all persons registered Republican}\}$$

Give the *rule* for deciding whether or not an element belongs in the following sets. (Shading with a pencil will help.)

a. D

b. G

c. H

d. $G \cup H$

e. $D \cap B$

f. $E \cup F$

g. $\tilde{A}$

h. $\tilde{H}$

i. $A \cup (B \cap C)$

j. $(A \cup B) \cap C$

Counting: Permutations and Combinations

Introduction

Set concepts become more helpful in probability problems if you learn to calculate the number of elements in sets. The small-case letter "n" is used as a shorthand symbol for "the number of elements in . . . ." For example, the sentence

$$n(A) = 5$$

is read, "The number of elements in set A is five." The sentence

$$n(A \cup B) = 20$$

is read, "The number of elements in the union of sets A and B is 20."

Several types of problems are discussed in this Chapter. These include methods for determining the number of elements in combinations of sets, the number of elements in a set of all possible *orderings* of the elements of a given set, the number of elements in the set of all possible *combinations* of elements in a given set, and special counting techniques. Chapter 17 will show the application of these "counting procedures" to the calculation of probabilities.

Factorials

It is helpful to digress at this point and review the arithmetic of factorials. (This topic was discussed in Chapter 8.) The major statistical application of the factorial operation is in the calculation of probabilities and in certain counting problems.

Recall that the factorial operator (!) defines a repeated multiplication operation. For example,

$$6! = 6 \cdot 5 \cdot 4 \cdot 3 \cdot 2 \cdot 1 = 720$$

and, in general,

$$n! = n(n-1)(n-2) \cdots 3 \cdot 2 \cdot 1$$

Also, recall that $1! = 1$ and $0! = 1$.

Try this set of problems. If you have trouble, go back to Chapter 8 and review the material on factorials.

PROBLEM 16.1

    a. $4! = ?$

    b. $7! = ?$

    c. $8! = ?$ [*Hint.* $8! = 8(7!)$]

    d. $10! \div 8! = ?$ (*Hint.* Use cancellation. Only two numbers need to be multiplied.)

    e. $(5!)(9!)/(4!)(7!) = ?$

    f. $(15-12)! = ?$

    g. What is $(n-2)!$ if $n = 7$?

Number of Elements in Combinations of Sets

DISJOINT SETS

Consider a simple question: "How many persons are in a sample that contains 50 men and 40 women?" The answer is obviously 90. However, the question illustrates a basic principle of disjoint sets: if two sets are disjoint (or mutually exclusive), then the number of elements in the union of the two sets equals the sum of the numbers of elements in the two sets.

The set of all male subjects and the set of all female subjects are certainly mutually exclusive. Therefore, the number of elements in the set of all subjects (90) is the simple sum of the numbers of elements in the set of male subjects (50) and the set of female subjects (40).

Symbolically, let M be the set of male subjects and F be the set of female subjects. We have

$$n(M) = 50 \qquad \text{and} \qquad n(F) = 40$$

Since M and F are mutually exclusive,

$$n(M \cup F) = n(M) + n(F) = 90$$

OVERLAPPING SETS

The problem is not quite as simple with pairs of sets that are not disjoint. Two nondisjoint sets have elements in their intersection that belong to both sets. The elements in the intersection should be counted once and only once if the sum desired is the number of elements in the set union.

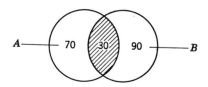

Figure 16.1

Let A and B be sets such that $n(A) = 100$ and $n(B) = 120$, but $n(A \cap B) = 30$. The value of $n(A \cup B)$ is desired. Figure 16.1 will help. $A \cap B$ is shown by shading.

The examination of Figure 16.1 shows clearly that the desired answer is

$$n(A \cup B) = 190$$

This value is obtained by summing the numbers of elements in the three disjoint sections of Figure 16.1.

Note that

$$n(A) + n(B) \neq 190$$

since $n(A) = 100$ and $n(B) = 120$. Algebraically, the value of $n(A \cup B)$ is given by

$$n(A \cup B) = n(A) + n(B) - n(A \cap B)$$

The figure shows how the sum $n(A) + n(B)$ adds the number 30 *twice*. By subtracting it *once*, we get the proper solution.

PROBLEM 16.2

		R Republicans	D Democrats	Total
L	Liberals	50	200	250
C	Conservatives	100	100	200
		150	300	450

The above table gives a hypothetical breakdown of Congress by party and by liberal-conservative belief. Find the following values by inspecting the table. Verify answers *d* and *f* by the algebraic formula.

a. $n(R)$

b. $n(L)$

c. $n(R) + n(L)$

d. $n(R \cup L)$

e. $n(C \cap D)$

f. $n(C \cup D)$

MORE THAN TWO OVERLAPPING SETS

In problems involving several overlapping sets, formulas can be found in some books on probability or set algebra. However, the problems can be directly attacked by overlapping circle diagrams or other systematic arrangements of the data.

PROBLEM 16.3

Consider Problem 3 at the end of the previous chapter. Suppose the following are true (where numbers are millions):

$$n(D) = 1$$
$$n(E) = 4$$
$$n(F) = 54$$
$$n(G) = 2$$
$$n(H) = 2$$
$$n(I) = 40$$
$$n(J) = 36$$

Determine the following by inspection and by formula if possible.

a. $n(B)$

b. $n(C)$

c. $n(B \cup C)$

d. $n(A)$

e. $n(A \cup B)$

f. $n(A \cup (B \cup C))$

Permutations

The word "permutation" means basically "order of occurrence." The meaning is best understood by example. Consider a race run by three contestants numbered from one to three. Suppose you are asked to bet on the final outcome of the entire race—that is, you are to guess who will win, who will be second, and who will lose. Before you should be willing to put up money on the bet, you should know in how many ways the race could wind up.

You might proceed by *enumeration*, that is, by listing all possible finish orders. Let's list these in a systematic way as follows.

	First Place	Second Place	Third Place
1.	1	2	3
2.	1	3	2
3.	2	1	3
4.	2	3	1
5.	3	1	2
6.	3	2	1

Thus, there are *six* possible finish orders in a three-person race. Each of these is a different *permutation* of the finish order. A bettor should seek five-to-one odds on the bet, assuming that the contestants are of about the same ability.

A quicker way to determine the total number of outcomes in the three-contestant race is by using the factorial operator. The solution is 3! or $3 \cdot 2 \cdot 1 = 6$.

If the race had *ten* contestants instead of three, it would be almost impossible to determine the number of possible permutations by enumeration without making a mistake and, in addition, many working days would be required for the task. However, the number of permutations can be determined relatively quickly by the factorial method: the answer is 10! or 3,628,800.

Rather than blindly accepting the fact that the factorial operation gives the proper number of permutations, let's systematically go through another example that might give some insight into the relationship of the factorial to the number of permutations.

Consider an investigation in which an interviewer wished to ask persons four questions that we shall symbolically designate as *a*, *b*, *c*, and *d*. Suppose the investigator knows that the response to any question might affect the response to subsequent questions. To prevent this "order effect" from biasing his interview results, he does not wish to present his questions in the same order to all interviewees. How many different orders (permutations) are possible?

The researcher has *four* choices for the first question. Suppose question *a* is chosen first. Then, the second question can be any of the *three* remaining (suppose he chooses *d*). The third question can be chosen in only *two* ways (since *b* and *c* are all that are left). The remaining question must be last as it is the only *one* left. We see the pattern emerge—4 × 3 × 2 × 1 = 24.

The pattern can also be seen if the researcher enumerates the 24 permutations systematically. The total set of 24 permutations is:

a b c d	b a c d	c a b d	d a b c
a b d c	b a d c	c a d b	d a c b
a c b d	b c a d	c b a d	d b a c
a c d b	b c d a	c b d a	d b c a
a d b c	b d a c	c d a b	d c a b
a d c b	b d c a	c d b a	d c b a

There are *four* subsets corresponding to the four possible first letters. For *each* of these four subsets, there are *three* subsets corresponding to the three possible second letters (these are separated by the underlines). This gives 4 × 3 = 12 subsets. *Each* of these 12 subsets has *two* possible third letters to give 4 × 3 × 2 = 24 total permutations. The final letter is determined by the first three, so we have 4 × 3 × 2 × 1 = 4! = 24 elements in the set of all possible permutations.

Subsets of Permutations

Those of you who are familiar with horse racing know that the bettor seldom bets on the total outcome of the race. One generally bets on the first three places only. Suppose there is a five-horse race. One usually bets for a horse to "win, place, *or* show," that is, to come in first, second, *or* third. Let's suppose, however, that we are asked to pick the *winner*, the *second* place horse, *and* the *third* place horse. The question is "In how many ways can this be done?" The answer is certainly not 5!, because 5! is the total number of outcomes including fourth and fifth places. The answer to our question is 5 × 4 × 3 = 60. There are *five* horses that can win. After choosing a winner, there are *four* horses that can come in second. After choosing the first two places, there are *three* remaining horses to choose for third place.

A similar example comes from a standard type of essay exam. A professor might list six questions and ask you to answer any three. You would probably choose the three that you think you can best answer.

Combinations of Events

If a person is to perform a set of tasks and each of these tasks can be accomplished in several ways, there is a simple way to determine the total number of possible "strategies" that can be taken. The total is the product of the number of alternatives for each task.

Travel provides a good example. Suppose there are two routes between cities A and B and four routes between cities B and C. A traveler who wishes to go from city A to city C has 2 × 4 or 8 possible routes.

A rat running a T-maze has two alternatives at each intersection of the maze. Suppose he must pass through four intersections of the maze. Then the rat has 2 × 2 × 2 × 2 or 16 choices for a route through the maze.

Suppose an examiner is evaluating patients with a personality battery that has three scales—neurosis, psychosis, and intro-extroversion. Each scale yields ten possible scores. The possible number of personality profiles that can be obtained with the test battery is 10 × 10 × 10 or 1000.

A slightly more complex example will show an extension of the general rule illustrated by the previous three examples. Suppose a car buyer wishes to buy a car from one of three manufacturers—Ford, Chevrolet, and Plymouth. Each manufacturer has a different number of models in its line (such as Fairlane, Mustang, Thunderbird, etc.) and let's assume that each model has the same number of styles. How many choices does the buyer have if the following table is correct for a given year?

	Models	Styles
Chevrolet	8	4
Ford	10	6
Plymouth	5	3

The answer is 8 × 4 + 10 × 6 + 5 × 3 = 107. If each model had a different number of styles, then the problem would be only slightly more complex and would be solved in a similar manner.

Numbers of Combinations

Suppose you are chairman of a six-man committee. You wish to choose a *three*-person subcommittee. How many possible committees (or *combinations*) can be formed (assuming that you are not on the subcommittee)? Name the five members *a*, *b*, *c*, *d*, and *e*. The possible committees are enumerated below.

a b c	a d e
a b d	b c d
a b e	b c e
a c d	b d e
a c e	c d e

Thus, ten committees of three persons can be chosen from the five available persons. We say, "There are ten *combinations* of five things taken three at a time."

Note the difference between a *combination* and a permutation. Committees {*a*, *b*, *c*} and {*a*, *c*, *b*} are both the same combination, but they are *not* the same *permutation*. The two groups, {*a*, *b*, *c*} and {*a*, *c*, *b*}, are *two different permutations* of the same combination. The concept "permutation" deals with *order* and the concept "combination" does *not* deal with order.

Now let's generate a method of calculating the number of committees that does not require complete enumeration. We can start by using the same technique that was used for developing the permutation concept.

First, choose a committee member. There are *five* ways to choose the first member. Suppose we choose "*b*." This leaves *a*, *c*, *d*, and *e*. We now have *four* choices for the second member. Suppose we choose *e*. Our third member must be chosen from the remaining *three* persons—*a*, *c*, and *d*.

The product $5 \cdot 4 \cdot 3 = 60$ is the number of permutations of three objects selected from five—but we are *not* interested in all of the permutations. If we choose *a* as the last subcommittee member, our subcommittee is {b, c, a}. This is the *same* committee (or *combination*) as {c, b, a} or {a, b, c}. In fact, there are *3! or six different permutations* of *each* possible triad of persons selected. That is, the value we want is *one-sixth of sixty*. By formula, we have

$$\frac{5!}{2!\,3!} = 10$$

since $5 \cdot 4 \cdot 3 = 5!/2!$.

The symbol $\binom{5}{3}$ is used to designate the number of *combinations* of five things taken three at a time. We have

$$\binom{5}{3} = \frac{5!}{3!\,2!} = 10$$

Note that $3! = (5 - 2)!$ and $2! = (5 - 3)!$ Here are some other examples.

a. $\binom{6}{4} = \frac{6!}{4!\,2!} = \frac{6 \cdot 5 \cdot 4!}{2!\,4!} = \frac{6 \cdot 5}{2}$

b. $\binom{10}{9} = \frac{10!}{1!\,9!}$

c. $\binom{7}{2} = \frac{7!}{2!\,5!}$

d. $\binom{7}{5} = \frac{7!}{5!\,2!}$

Finish the calculations in the examples. You should get a, 15; b, 10; c, 21; and d, 21.

Note that $\binom{7}{2} = \binom{7}{5}$

In general, the formula for the number of combinations of n things taken r at a time is

$$\binom{n}{r} = \frac{n!}{r!\,(n - r)!}$$

Exercise. Verify that $\binom{n}{r} = \binom{n}{n - r}$ by writing each side out in factorial form.

Perhaps the symbol "0!" can make more sense now. If you have a bag containing five pieces of candy, in how many ways can you remove exactly *no pieces*? By formula, we have

$$\binom{5}{0} = \frac{5!}{0!\,5!} = 1$$

There is *one* way to remove *no* items—that is, by leaving the candy alone!

Also, for consistency, consider the ways to remove *five* pieces of candy from a bag containing *five* pieces. There is obviously one way to do it— namely, to take all of it. But the formula should work in this case, too. We have

$$\binom{5}{5} = \frac{5!}{5!\,0!}$$

We must make 0! equal to *one* in order that $(5!)/(5!)(0!)$ will be *one*, which is obviously the correct solution.

Let's take some fairly practical examples.

PROBLEM 16.4

a. There are 25 rats in a psychologist's "rat colony." How many possible random samples of 20 rats can be taken from the colony? Express the answer in factorial form, expand the factorials, and cancel as many terms as possible before multiplying or dividing.

b. A football coach has 12 players trying out for line positions (where only 7 can play at once) and 10 players trying out for backfield positions (where only 4 can play at once). How many different teams are possible (assuming that the coach does not switch backs and linemen)? *Hint.* Find the number of linemen combinations and the number of backfield combinations as the first step.

c. An experimenter wishes to make up three-letter nonsense syllables using *only one vowel* in each. Assume that "y" is used only as a consonant. How many three-letter syllables are possible?

The Binomial Expansion

The combination formulas provide a technique that is very helpful in a wide variety of statistical problems based on the binomial expansion. The quadratic expression $(a + b)^2$ should be quite familiar to the reader. The expansion is

$$(a + b)^2 = a^2 + 2ab + b^2$$

However, the expansion of the same expression to a higher power, say the fourth power, is a different matter:

$$(a + b)^4 = ?$$

The terms of the expansion are all of a general form involving the combinational formulas from the previous section. Using the combination symbols, we can expand $(a + b)^4$ quite systematically:

$$(a + b)^4 = \binom{4}{0}a^4b^0 + \binom{4}{1}a^3b^1 + \binom{4}{2}a^2b^2 + \binom{4}{3}a^1b^3 + \binom{4}{4}a^0b^4$$

Remember that anything raised to the zero power is one and that $0! = 1$. Simplify the expansion. You should get

$$(a + b)^4 = a^4 + 4a^3b + 6a^2b^2 + 4ab^3 + b^4$$

The discussion of the importance of this section in statistical applications must wait for the probability chapter. The reader should focus on the use of the combination symbolization for determining the coefficients of the expansion.

The coefficients, called *binomial coefficients* or *binomial numbers*, are, as you learned previously, the numbers of *combinations* of things. Let's look again at the binomial coefficients for $(a + b)^4$. They are, in order, 1, 4, 6, 4, and 1. By definition, these numbers are:

1. $\binom{4}{0} = 1$—the number of ways of taking *no* things out of 4

2. $\binom{4}{1} = 4$—the number of ways of taking *1* thing out of 4

3. $\binom{4}{2} = 6$—the number of ways of taking *2* things out of 4

4. $\binom{4}{3} = 4$—the number of ways of taking *3* things out of 4

5. $\binom{4}{4} = 1$—the number of ways of taking *4* things out of 4

So the combinational formulas are seen to be a quick way to find the binomial coefficients. These coefficients can be quickly tabulated without even doing the factorial multiplications. At least they can be tabulated into a table that is reasonably useful.

The tabulation procedure was developed by Pascal, one of the great mathematicians. The resulting table bears his name and is known as Pascal's Triangle. The table is built upon one of Pascal's discoveries:

$$\binom{n}{r} + \binom{n}{r + 1} = \binom{n + 1}{r + 1}$$

where n and r are any numbers with the condition that r is not larger than n. You can verify the formula by algebraic manipulation or by trying out a few examples.

Pascal's Triangle can be easily constructed. Take a sheet of paper and write two "1's" in the center of the first line. You should have this figure:

1. 1 1

Now, on the second line write "1" to the left of the first "1" on line one. Add the two digits on line one and place the sum (2) between the two "1's." Finally, place a one to the right of the sum. You should have the following.

1. 1 1

2. 1 2 1

Continue building the triangle by beginning and ending with a "1" and every other numeral will be the sum of the two above it. Look at the following table and fill in the missing values.

1.   1 1

2.   1 2 1

3.   1 3 3 1

4.   1 4 6 4 1

5. 1 5 – – – 1

You should have entered 10, 10, and 5 in line 5. Do two more lines. If you do them correctly, the final line should be 1, 7, 21, 35, 35, 21, 7, and 1. Notice that line 4 gives the binomial coefficients for the expansion of $(a + b)^4$. Line 7 gives the binomial coefficients for the expansion of $(a + b)^7$.

Moreover, the table gives the values of the combinations of n things taken r at a time. Line 7 gives the following values (in order):

$$\binom{7}{0} = 1$$

$$\binom{7}{1} = 7$$

$$\binom{7}{2} = 21$$

$$\binom{7}{2} = 35$$

PROBLEM 16.5

Use the table to determine these numbers.

a. $\binom{7}{5}$

b. $\binom{3}{1}$

c. $\binom{5}{3}$

You might wish to verify these values by calculation.

Problems for Review

1. Perform the indicated multiplications.

a. $5!$

b. $8!$

c. $1!$

d. $0!$

e. $\dbinom{4}{2}$

f. $\dbinom{6}{4}$

g. $\dbinom{14}{11}$

h. $\dbinom{7}{2}$

i. $\dbinom{8}{0}$

j. $\dbinom{13}{13}$

k. $\dbinom{7}{5}$

l. $\dbinom{6}{2}$

2. In a particular high school there are 74 male 10th graders, 86 female 10th graders, 140 male 11th graders, 95 female 11th graders, 110 male seniors and 115 female seniors.

In this school, how many *elements* are in the following sets?

a. {all males}
b. {all seniors}
c. {all females}
d. {all nonseniors}
e. {male sophomores} $\cup$ {all males}
f. {all females} $\cap$ {all sophomores}
g. {all males} $\cap$ {all nonseniors}
h. {all sophomores} $\cup$ {all senior girls}

3. a. In Problem 2, which problems involved combining only mutually exclusive sets?
   b. In Problem 2, which problems involved combining *some* sets that are *not* mutually exclusive?
   c. In Problem 2, which problems involved combining sets some of which were mutually exclusive and some of which were not?

4. Draw Venn diagrams for Problems 2e and 2f if you have not already done so. If you solved Problem 2 by Venn diagrams, be sure you can also do it by formulas.

5. Suppose an English professor lists for his class six anthologies and twelve paperbacks. He asks each class member to purchase two of each. How many *different* sets of textual materials are possible with these instructions?

6. An experimenter wishes to study the effect of three diets. He has 20 volunteers from which he will choose 18. Each chosen participant will use

each of the three diets for one month. Appropriate measurements will be made periodically. He wishes to avoid "order effect" by using all possible orders of dieting.

a. How many orders are possible?
b. How many different groups of participants are possible?
c. How many participants should diet in any one particular order of diets, assuming the experimenter wishes to "balance" the experiment with respect to an order effect?

17

Elementary Probability

Introduction

Probability concepts are the foundation of statistics. Most statistics courses will contain some discussion of probability, and many courses will deal heavily with probability topics.

However, it is possible for a course in statistics to have no special unit on probability if the course is highly "cookbook" in orientation. The omission of probability training usually is considered a mistake because the application of probability ideas is necessary in any kind of statistical endeavor beyond the most simple description techniques. If you receive an inadequate background in elementary probability, then you will suffer in the *interpretation* of statistics, and it is in *interpreting* statistics that we all wish to be most skillful.

This chapter deals only with the elements of probability theory and applications. Most applied statistics courses will repeat a great deal of this content and will not add too much of a more complex nature. By getting an introduction now, perhaps the unit in your statistics class will be more meaningful. Also, it is hoped that if all students have been previously introduced to probability ideas, then the statistics instructor can provide a greater understanding of the probability foundations of statistics than is possible if he must begin at the beginning.

Something should be said about examples. Many fundamental probability notions are best explained by gambling examples—coin tossing, dice throws, cards, and the like. These examples will be used as needed, but an attempt will always be made to choose examples from practical statistical problems.

The Probability Language

THE BASIC PROBABILITY DEFINITIONS

It is convenient to talk in terms of an experiment. Here, the word "experiment" will be used loosely to refer to any act under the control of an experimenter in which he is not sure what will happen. However, the experimenter usually knows all of the elements of the set of outcomes that is possible.

The experiment might be as simple as the tossing of a coin—the coin may fall "heads" or "tails." The set of outcomes has only two elements—"heads" and "tails." One does not know until the toss is completed whether the outcome will be a head or a tail. The words *"event"* and *"outcome"* are used to mean the results of the experiment.

The experiment can be something as highly complex as the estimation of the average intelligence of a randomly selected group of subjects. Here, the experiment is the complex act of testing, scoring, and computing that leads to a particular observed numerical average (the outcome). This experiment would result in one possible number out of an infinitely large set of outcomes.

The concept of *probability* can best be developed by examples that are neither as simple as the first example nor as complex as the second. A card-drawing experiment will provide ample examples. Let the experiment be the act of drawing a card from a well-shuffled standard fifty-two card deck. What are the possible outcomes? There are four suits of thirteen cards each. There are four cards of each of the thirteen "kinds" (ace, deuce, etc.). There are two colors of twenty-six cards each. However, these possibilities are all special cases of the set of possible outcomes. There are obviously fifty-two cards in the set of possible outcomes.

Several questions can be asked. (a) What is the probability of drawing a red card? (b) What is the probability of drawing a spade? (c) What is the probability of drawing a king?

These questions can be answered by considering the common properties of a probability measure. A probability measure is, first of all, a number that is zero, one, or a fraction in between zero and one.

The *lowest value is zero* and this value is reserved for the probability of an *impossible event*. For example, what is the probability of drawing the fifteen of spades? The answer is zero because the event is impossible. Stated formally, the desired event is not an element of the set of possible outcomes.

The *highest probability value is one*. This value is reserved for the *certain event*. For example, what is the probability that a die will show less than 10 spots? The answer is *one* because *all* possible outcomes have the desired property of being less than 10.

Events that are neither impossible nor certain are assigned a probability measure between zero and one. If you have no evidence that leads you to believe that an event is more likely to occur than it is not to occur, then it is logical to choose the number $\frac{1}{2}$ to be the probability of the event. If you say that "the probability of tossing a head is $\frac{1}{2}$," you mean that you have no evidence that the coin is biased in favor of heads or in favor of tails. You believe that a head and a tail are equally likely outcomes of the coin-tossing experiment. If you believe (perhaps from past experience) that the coin is biased in favor of "heads," then you would use a number between one-half and one to be the probability measure.

There is one way of deciding what number to use as the probability of an event that is applicable in many statistical problems. If the experiment gives rise to a set of equally likely outcomes, then you can use the ratio of the number of outcomes corresponding to the event of interest to the total number of outcomes. Although, this statement is often called the "classical definition" of probability, it is best considered as a way of calculating probability measures in the special case in which all outcomes are equally likely.

We can use the "classical definition" to solve the three-card problems.

(a) What is the probability of drawing a red card? There are 26 red cards; that is, 26 outcomes correspond to the event "drawing a red card." There are 52 total possible equally likely outcomes. Therefore, the probability of drawing a red card is $\frac{26}{52}$ or $\frac{1}{2}$.

(b) What is the probability of drawing a spade? There are 13 spades; that is, there are 13 outcomes corresponding to the event "drawing a spade." Therefore, the probability of getting a spade is $\frac{13}{52}$, or $\frac{1}{4}$.

(c) What is the probability of drawing a king? Since there are 4 kings, the probability of drawing a king is $\frac{4}{52}$, or $\frac{1}{13}$.

Now some symbolism will be introduced that will tie this section to the set theory chapters. Let T be the set of all cards $[n(T) = 52]$. Let R be the set of all red cards $[n(R) = 26]$. Let S be the set of all spades $[n(S) = 13]$. Let K be the set of all kings $[n(K) = 4]$. The symbol "$P\{X\}$" will be used for "the probability of X." Thus, "$P\{K\} = \frac{1}{13}$" is read, "The probability of drawing a king is $\frac{1}{13}$," and the statement "$P\{S\} = \frac{1}{4}$" means, "The probability of drawing a spade equals $\frac{1}{4}$."

We can use set notation in the estimation of probabilities. Let T be the total set of outcomes of an experiment. There are $n(T)$ elements in T. Let A be a set of outcomes (a subset of T) with $n(A)$ elements. Then the probability of occurrence of an event which is an element of A can be estimated by

$$P\{A\} = \frac{n(A)}{n(T)}$$

PROBLEMS

17.1. A card is drawn at random from a well-shuffled standard deck. What is the probability that it is:

a. A black card?
b. A face card?
c. The four of clubs?

17.2. One die is tossed. What is the probability that the die will show:

a. Six spots?
b. An even number of spots?
c. An odd number of spots?

THE PROBABILITY OF NONEQUIPROBABLE EVENTS

The examples in the preceding section all were examples of *equally likely* or *equiprobable events*. In the drawing of a card from a standard deck, each of the 52 cards has the same chance (namely, $\frac{1}{52}$) of being chosen. In the tossing of a "fair" coin, both "head" and "tail" have an equal chance ($\frac{1}{2}$) of occurring. In the tossing of a "fair" die, any of the six sides has an equal chance ($\frac{1}{6}$) of showing.

Many statistical problems involve equiprobable events. The most common is random sampling. If there exist 100 persons in a population of interest and an experimenter wishes to choose one of these persons by random sampling, then the person is chosen in a way that each of the 100 persons has a probability of being chosen equal to $\frac{1}{100}$.

However, more often than not, the events of interest are *not* equally probable. Consider the simple experiment of throwing two dice at once and counting the sum of the dots that show. There is only one way to get a "two" and that is if both dice show exactly one spot. On the other hand, there are several ways to get a "seven." These include a "six" and a "one," and a "five" and a "two," as examples. In fact, the probability of rolling a "seven" is seven times as great as that of rolling a "two."

Consider the problems of measuring persons. If one person is chosen at random, the probability that his IQ is between 100 and 105 is much greater than the probability that his IQ is between 125 and 130. The probability that the person is in the lower socioeconomic bracket is far greater than the probability that he is in the upper socioeconomic bracket.

In experiments involving nonequiprobable events the classical rule is inappropriate. A helpful rule is the "relative frequency" rule for estimating probability measures. This rule says to estimate the probability of an event by repeating the experiment many times. The ratio of the number of times a desired event occurs to the number of experiments performed is the "relative frequency" estimate of the probability of the event.

For example, if the experiment is coin tossing and "heads" occurs 60 times in 100 tosses, then the "relative frequency" estimate of the probability of a head is $\frac{60}{100}$, or .6. Using the "classical" rule, you would say that the probability of a head is $\frac{1}{2}$. The "relative frequency" value of .6 is based on experimental evidence.

PROBLEMS

Suppose you are interested in betting on a horse that is one of six horses in a race.

17.3. What is the classical probability estimate of your horse winning?

17.4. How would you use the "relative frequency" concept to decide on the probability of your horse winning?

THE PROBABILITY OF NONRECURRING EVENTS

Some phenomena do not lend themselves to either the classical or the relative frequency estimates of probability measures. Once-in-a-lifetime events are examples. What is the probability of rain tomorrow? Tomorrow only occurs once. What is the probability that you will have an automobile accident on your way home tonight? You will make this trip only once. However, these two examples (and others like them) do have reasonable answers based upon knowledge of general weather and traffic conditions. Usually, one can suggest a number that is a reasonable probability measure.

If this is the rainy season and it is now raining, you might say that the probability of rain tomorrow is .8, or .9, or even 1.0, depending upon your own personal opinion of the weather situation. Weathermen make such personal probability statements all the time. They have access to considerable weather information, but the final probability estimate is often highly subjective.

This concept of "subjective probability" or "personal probability" is an important theoretical development and is the basis of what is known as "Bayesian statistics." The idea is controversial and many statisticians dislike it. However, it is the only way to think about the probability of some non-recurring events and appears to be a highly useful idea in many statistical problems.

INDEPENDENT AND NONINDEPENDENT EVENTS

Two events are said to be *independent* if either event in no way affects the other. Coin tossing affords the easiest example. If two coins are fairly tossed, then the outcomes of the tosses are independent—one toss does not affect the other. If the probability of a head is $\frac{1}{2}$ on the first toss, whether or not a head occurs on the first toss, the probability of a head occurring on the second toss is also $\frac{1}{2}$.

Two events are *not independent* if one event affects the other. Card drawing will demonstrate this condition. In a fair card deck the probability of drawing a heart on the first draw is $\frac{13}{52}$. However, the probability of drawing a heart on the second draw is *not* $\frac{13}{52}$. The actual probability depends on what was drawn on the first draw. It is $\frac{12}{51}$ if the first card was a heart and it is $\frac{13}{51}$ if the first card was not a heart. (Of course, if one card is drawn and then replaced randomly in the deck *before* the second card is drawn, then the events would be independent.)

Random sampling is the best statistical example of independent events. Random sampling is *defined* as choosing experimental units or subjects in such a way that each choice is independent of all other choices.

MUTUALLY EXCLUSIVE AND NONMUTUALLY EXCLUSIVE OUTCOMES

Think of a simple experiment that has two outcomes. The two outcomes are mutually exclusive if they both cannot happen simultaneously. The simple experiment of tossing a single coin results in two mutually exclusive outcomes—*either* a head *or* a tail can occur, but *both cannot occur simultaneously*.

A counter-example can be constructed from the experiment of drawing one card from a standard deck of cards. A card can be both an ace and a spade—so the outcome of drawing an ace and the outcome of drawing a spade are *not* mutually exclusive.

The term "mutually exclusive" can also be used in regard to sets of outcomes as well as particular outcomes. Two sets of possible outcomes to an experiment are *mutually exclusive* if the sets are *disjoint*—that is, if they do not overlap, or, similarly, if their intersection is an empty set.

A set of meaningful examples can be provided by considering the experiment of selecting a person at random from the world's population. Human beings can be classified or measured on an infinitely large number of characteristics, some of which are mutually exclusive sets and some of which are not. The researcher who deals with human subjects frequently spends considerable effort in trying to determine whether or not a measuring device or classification scheme yields mutually exclusive categories. The following are examples of questions that can be asked that imply mutually exclusive sets of outcomes.

1. Is the person male or is the person female? (The implied outcomes "male" and "female" are mutually exclusive sets.)

2. Was the person born in England or was he born in the U.S.A.? (The implied mutually exclusive outcomes can be merely "English-born" and "U.S.A. born" or the outcomes could be a listing of several or all possible birthplaces.)

3. If the person took a 10-item test, did he answer exactly 10 questions correctly or did he answer 9, or 8? (The number of outcomes implied in this set of outcomes depends on the number of items on the quiz. With 10 questions, there are 11 mutually exclusive score categories possible—including zero.)

The following are examples of questions that can be asked that do *not* imply mutually exclusive sets of outcomes.

1. Is the person male, or is the person English-born? Since the person can belong simultaneously to both sets—{all males} and {all English-born persons}—the outcomes of being male and being English-born are not mutually exclusive.

2. Does the person have high verbal intelligence or does he have high quantitative ability? Certainly some persons are high in both abilities and some persons are low in both.

The reader should make a careful distinction between the concepts "independent events" and "mutually exclusive outcomes." The latter concept refers to possible outcomes of an experiment. For example, one toss of a coin might yield the mutually exclusive outcomes of a head or of a tail. The concept of independence is irrelevant in this example.

On the other hand, if the experiment involves two tosses of a coin, we expect the *tosses* to be independent, and thus the event of observing the outcome on the first toss is *independent* of the event of observing the outcome on the second toss. The concept of "mutually exclusive outcomes" is irrelevant in the second example.

This second experiment does involve mutually exclusive outcomes which are two heads, two tails, a head on the first toss but not on the second, and a tail on the first toss but not on the second (H–H, T–T, H–T, and T–H). So the second experiment consists of two independent events and can possibly lead to any one of four mutually exclusive outcomes.

Calculating Probabilities

COMBINING MUTUALLY EXCLUSIVE OUTCOMES

The probability of occurrence of *any one* of a set of *mutually exclusive outcomes* is the *sum of the probabilities of each outcome*. If *A* and *B* are mutually exclusive outcomes, then the probability that *either A or B* will occur ($P\{A$ or $B\}$) is

$$P\{A \text{ or } B\} = P\{A\} + P\{B\}$$

For example, what is the probability that a card drawn at random from a well-shuffled, standard deck will be *either* a spade *or* a club? The answer is obviously one-half, since clubs plus spades account for one-half of the deck. By formula, however, we have the probability of a spade is $\frac{1}{4}$; the probability of a club is $\frac{1}{4}$; the two outcomes are mutually exclusive; therefore, since $\frac{1}{4} + \frac{1}{4} = \frac{1}{2}$, the probability of a club *or* a spade being drawn is $\frac{1}{2}$.

A similar example is, "What is the probability that a card drawn at random is an ace?" We know that four aces exist, so the answer is $\frac{4}{52}$ or $\frac{1}{13}$, but this can also be done by formula. The probability of getting the ace of spades is $\frac{1}{52}$; the probability of getting the ace of clubs is $\frac{1}{52}$; the probability of getting the ace of hearts is $\frac{1}{52}$; the probability of getting the ace of diamonds is $\frac{1}{52}$; these four outcomes are mutually exclusive; therefore, the answer is $\frac{1}{52} + \frac{1}{52} + \frac{1}{52} + \frac{1}{52} = \frac{4}{52}$.

PROBLEMS

17.5. What is the probability of drawing a card from a well-shuffled, standard deck that is:

a. *Either* a club, spade, *or* heart?
b. *Either* a club *or* a red ace?
c. *Either* a deuce, trey, *or* four?

17.6. Suppose 1000 lottery tickets are sold. The tickets are numbered from 1 to 1000. What is the probability that the winning number will be:

a. Either 1 or 1000?
b. Less than 100?
c. Less than 200 or greater than 900?

JOINT PROBABILITY OF INDEPENDENT EVENTS

The probability that independent events will occur jointly is the product of the probabilities of each event. If A, B, and C are independent events, then the probability that A, B, and C will occur ($P\{ABC\}$) is

$$P\{ABC\} = P\{A\}P\{B\}P\{C\}$$

Probably the simplest example of this formula is the calculation of the probability of tossing a coin twice and getting a head on *each* toss. If the probability of getting a head on the first toss is $\frac{1}{2}$, then the probability of getting a head on the second toss is also $\frac{1}{2}$. The two events are independent; therefore, since $\frac{1}{2} \cdot \frac{1}{2} = \frac{1}{4}$, the probability of getting two heads on two tosses is $\frac{1}{4}$.

Consider an example from achievement tests. Suppose you take a test consisting of 5 four-choice multiple-choice questions. What is the probability that you can correctly answer all of the questions by blind guessing? The assumption that guessing is "blind" rather than based on "educated guesses" allows us to assume that each guess is independent of other guesses. The solution is

$$\frac{1}{4} \times \frac{1}{4} \times \frac{1}{4} \times \frac{1}{4} \times \frac{1}{4} = (\frac{1}{4})^5 = \frac{1}{1024}$$

If you correctly answer all five questions, the instructor can be quite sure that you did not do it by guessing since the probability is so small.

Problem 17.7

Suppose you draw a card from a standard deck, look at it, replace it in the deck, reshuffle, and draw a second card. What is the probability that:

  a. Both cards are spades?
  b. Both cards are aces?
  c. The first card is a spade and the second card is an ace?
  d. Only the *first* card is an ace?
  e. Only the *second* card is an ace?

Conditional probability

The term *"conditional probability"* refers to those experiments in which the experimenter has some prior knowledge. That is, there exist certain *conditions* of which the experimenter is aware and which affect the probabilities involved.

Let's take a practical example. Suppose a student, John Smith, applies for admission to a college that has a 50% graduation rate. With no other information, the probability that Smith will graduate must be estimated by relative frequency to be $\frac{1}{2}$. However, suppose we know that Smith has a College Board verbal score of 800. Knowing this *condition* we can give a better estimate of the probability of his graduating. This new estimate will be based on the proportion of enrollees who score near 800 who have graduated in the past.

A card example previously used will serve not only to clarify the meaning of conditional probability but also to show the effect on probabilities of knowing certain existing conditions. Consider a game in which you win if a spade is drawn and lose if a spade is not drawn. The first player draws a card and it is not replaced. You play second. What are your chances of winning? Your draw is *not* independent of the first draw. If the first player has drawn a spade, then the probability of your doing so has decreased slightly since one spade is gone. If the first player failed to draw a spade, then the probability of your drawing a spade has increased slightly. In either case, there are 51 cards in the reduced deck. In one case there are 12 spades remaining ($P = \frac{12}{51}$) and in the other case there are still 13 spades ($P = \frac{13}{51}$). Notice that if the first player does *not* show you his card, this discussion is academic. You *must know* what his card is if you wish to calculate the new conditional probabilities.

A vertical line ($\mid$) is commonly used to denote conditional probability. The symbol combination

$$P\{\text{spade} \mid \text{spade on draw 1}\} = \tfrac{12}{51}$$

could be read as, "The probability of drawing a spade *given* that a spade was drawn by the first player is $\frac{12}{51}$." The words or symbols to the left of the vertical line tell what outcomes are desired. The words or symbols to the right of the vertical line tell what conditions are known.

The first example might be written symbolically as

$$P\{\text{Smith} \in \{\text{all future graduates}\} \mid \text{Smith} \in \{\text{all students scoring 800}\}\} = .85$$

The symbolic phrase is read, "The probability that Smith is an element in the set of all future graduates *given* that he is an element in the set of all students scoring 800 is .85."

The primary value in knowing preexisting conditions is that the total number of possible outcomes is reduced. This fact always leads to more accurate probability estimates. (The probability value might be higher or lower.) The job of a researcher essentially is to *determine conditions* so that more accurate probability estimates can be made.

Reconsider the two examples that were used to introduce the topic of conditional probability. In the card example, the denominator of the conditional probability is 51 since one card has been drawn. The numerator is 13 if no spade was drawn on the first draw, and it is 12 if a spade was drawn and therefore is missing.

The other example dealt with the prediction of graduation. Without knowledge of the college board scores, the probability of John Smith graduating could be estimated by data from the school's history. One estimate would be the ratio of the number of graduates to the number of persons who enrolled four years earlier. However, the knowledge of the test score means that the researcher can focus on only the persons scoring in the 800 range. The total number of enrollees who scored 800 would be the new denominator. The new numerator would be the total number of 800-scorers who eventually graduated.

In set language, the numerator is the number of elements in the intersection of the set of graduates with the set of 800-scorers, or, in "English," it is the number of persons who *both* graduated *and* scored 800. The denominator is the new base—the number of 800-scorers.

CALCULATING JOINT PROBABILITIES OF NONINDEPENDENT EVENTS AND CONDITIONAL PROBABILITIES

These two topics are combined because they are interdependent. We cannot calculate by formula the combined probability of nonindependent events A and B ($P\{AB\}$) without knowing either $P\{A \mid B\}$ or $P\{B \mid A\}$. On the other hand, we cannot calculate either $P\{A \mid B\}$ or $P\{B \mid A\}$ without knowing $P\{AB\}$. The basic formulas for joint probabilities are

$$P\{AB\} = P\{A \mid B\}P\{B\}$$

and, similarly,

$$P\{AB\} = P\{B \mid A\}P\{A\}$$

In other words, we have in the first formula, "The probability that both events A and B will occur is equal to the product of the probability that B will occur times the conditional probability that A will occur given that B does occur." The second expression can be written out similarly.

Now the dilemma arises when we ask how to find the conditional probabilities. The two formulas can be "solved" for the conditionals by standard algebraic manipulation to give

$$P\{A \mid B\} = \frac{P\{AB\}}{P\{B\}}$$

and

$$P\{B \mid A\} = \frac{P\{AB\}}{P\{A\}}$$

The dilemma posed by the interrelationship of joint probability and conditional probability is not the problem that it appears to be. The experimenter usually has the necessary information, although it is frequently determined empirically.

Consider the following example. It will show the interrelationship between conditional and joint probability. Suppose we know that 30% of all teenagers in a certain city slum area are delinquents. Suppose we also know that in this city 40% of all teenagers live in this slum area. We have, by counting, also determined that the percentage of the entire city teenage population that is *both* delinquent *and* residing in this slum area is 12%. Using D for "delinquent" and S for "slum resident," we have

$$P\{D \mid S\} = \frac{P\{DS\}}{P\{S\}} = \frac{.12}{.4} = .3$$

$$P\{DS\} = P\{D \mid S\}P\{S\} = (.3)(.4) = .12$$

In this example, we knew that the three values $P\{S\}$, $P\{DS\}$, and $P\{D \mid S\}$ were .4, .12, and .3, respectively. We needed to know only $P\{S\}$ and *either* $P\{D \mid S\}$ *or* $P\{DS\}$ but not both since the third value can be obtained by formula.

What is the probability that a New York citizen, chosen at random, is a liberal (L) *and* Republican (R)? You can determine the probability empirically by determining the number of New York Republicans and from this number finding out how many are liberals. Or, you might first determine the number of liberals, and from these liberals, determine the number of Republicans. The first method would provide base data for using the formula

$$P\{LR\} = P\{L \mid R\}P\{R\}$$

while the second method would provide the data for using the formula

$$P\{LR\} = P\{R \mid L\}P\{L\}$$

Of course, we could avoid the formula by determining the membership size of the four mutually exclusive sets $L \cap R$, $L \cap \tilde{R}$, $\tilde{L} \cap R$, and $\tilde{L} \cap \tilde{R}$. The value $n(L \cap R)$ would directly give the empirical estimate of $P\{LR\}$.

There is a very important special case of the formula

$$P\{AB\} = P\{A \mid B\}P\{B\} = P\{B \mid A\}P\{A\}$$

This is the case of independence of A and B, which was discussed previously. When A and B are independent, then A has no effect on B and B has no effect on A. Stated mathematically, if A and B are independent, then

$$P\{A \mid B\} = P\{A\}$$

and

$$P\{B \mid A\} = P\{B\}$$

Substituting these values into the formula for $P\{AB\}$ gives

$$P\{AB\} = P\{A\}P\{B\}$$

the result that was given in the earlier section.

COMBINING OUTCOMES THAT ARE NOT MUTUALLY EXCLUSIVE

If two outcomes, A and B, are *not* mutually exclusive, then the probability of *either A or B* occurring is *not* $P\{A\} + P\{B\}$. The correct probability also

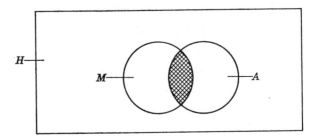

Figure 17.1

involves the probability that both outcomes happen together, which is a possibility with nonmutually exclusive outcomes.

Consider this example. What is the probability that a human being selected at random is *either* a male *or* an American? Let's use a Venn diagram (Figure 17.1). Let the rectangle H represent the set {all humans}. Circle M represents the set {all males}, and circle A represents the set {all Americans}. The sets are *not disjoint,* so the outcome "male" and the outcome "American" are *not* mutually exclusive. The shaded area is $M \cap A$.

The desired probability is

$$P\{M \text{ or } A\} = \frac{n(M \cup A)}{n(H)}$$

which can be expanded to

$$P\{M \text{ or } A\} = \frac{n(M) + n(A) - n(M \cap A)}{n(H)}$$

This can be rewritten as

$$P\{M \text{ or } A\} = \frac{n(M)}{n(H)} + \frac{n(A)}{n(H)} - \frac{n(M \cap A)}{n(H)}$$

which is

$$P\{M \text{ or } A\} = P\{M\} + P\{A\} - P\{MA\}$$

In general, if A and B are nonmutually exclusive outcomes of an experiment, then the probability that *either A or B* will occur is

$$P\{A \text{ or } B\} = P\{A\} + P\{B\} - P\{AB\}$$

This formula can also be applied in the special case in which A and B are mutually exclusive. In this case, $P\{AB\} = 0$ and the formula reduces to

$$P\{A \text{ or } B\} = P\{A\} + P\{B\}$$

which is the formula given in an earlier section.

PROBLEM 17.8

One card is drawn from a standard, well-shuffled deck. What is the probability that it is a:

   a. Spade *or* a club?
   b. Spade *or* an ace?
   c. Face card *or* spade?
   d. Five, six, seven, eight, *or* spade?
   e. Black card *or* a face card?

Bayes' Theorem

Bayes, an early student of probability theory, studied the relationships among conditional probability formulas. His work, especially the set of formulas now known as Bayes' Theorem, used with the concept "personal probability," is the basis of an approach to statistics that is becoming increasingly important and widespread. This approach, appropriately called "Bayesian statistics," is as much a philosophy of decision making as a method of calculation.

The statistical theory will not be discussed. However, an introduction to Bayes' formula will assist you in the study of Bayesian statistics if your statistics course moves in this direction. The procedures can be generalized to be applicable to very complex problems. However, we shall focus only on the simplest form of Bayes' Theorem.

The starting point is the observation that two ways exist to write $P\{AB\}$ in terms of conditional probability. You will recall that these are

$$P\{AB\} = P\{A \mid B\}P\{B\}$$

and

$$P\{AB\} = P\{B \mid A\}P\{A\}$$

So, we can write as a true statement

$$P\{A \mid B\}P\{B\} = P\{B \mid A\}P\{A\}$$

Dividing both sides of this equation by $P\{B\}$ gives the simplest version of Bayes' Theorem:

$$P\{A \mid B\} = \frac{P\{B \mid A\}P\{A\}}{P\{B\}}$$

An elaboration of the denominator puts the formula into a more useful form. First consider the equation

$$P\{B\} = P\{AB\} + P\{\tilde{A}B\}$$

(This statement is true, and if its truth is not obvious to you, think about it until it is clear.) This form of $P\{B\}$ can be further modified by using the conditional probability formulas for $P\{AB\}$ and $P\{\tilde{A}B\}$.

We have, finally, a very useful version of Bayes' Theorem:

$$P\{A \mid B\} = \frac{P\{B \mid A\}P\{A\}}{P\{B \mid A\}P\{A\} + P\{B \mid \tilde{A}\}P\{\tilde{A}\}}$$

Bayes' formula is used quite effectively in selection and diagnosis-type research. These research examples typically are characterized by knowledge of results from prior study, and this *prior knowledge* is used to estimate the *conditional* probabilities through Bayes' formula. (This is where "personal probability" comes in.)

A typical example will show the use of Bayes' formula. The example is fairly long, but is a complete application of the Bayesian approach. Suppose the Rorschach Test (inkblots) has been given to a large group of hospitalized psychotics and to a large group of "normal" subjects. Suppose the Rorschach is scored for confused thinking—that is, subjects are scored as "highly confused" (C) or not confused $(\tilde{C})$. Several questions can be asked in regard to the *future* use of the Rorschach for screening mental health cases as non-normal $(\tilde{N})$ or normal (N). These include:

1. Given that a person scores confused (C), what is the probability that he is psychotic $(\tilde{N})$? (What is $P\{\tilde{N} \mid C\}$?)
2. Given that a person scores nonconfused $(\tilde{C})$, what is the probability that he is normal (N)? (What is $P\{N \mid \tilde{C}\}$?)
3. In either case, what is the probability of a correct judgment? (What is $P\{\tilde{N}C\} + P\{N\tilde{C}\}$?)

Hypothetical results from prior research are as follows.

1. 80% of psychotics score C.
2. 20% of normals score C.
3. 10% of all persons are psychotic. (A wild overestimate, but satisfactory for tutorial purposes!)

The three bits of information may be combined to give 26% as the estimate of the number of *all* persons who score C.

From these data we can *estimate* the following probabilities:

$$P\{\tilde{N}\} = .10 \qquad P\{C \mid \tilde{N}\} = .80$$
$$P\{C\} = .26 \qquad P\{C \mid N\} = .20$$

Note that the data give the proportion of psychotics who score C, *not* the proportion of C-scorers who are psychotic. It is the latter figure that is important ($P\{\tilde{N} \mid C\}$).

Let's find the desired values by substituting the data into the general form of Bayes' Theorem.

1. $P\{\tilde{N}\,|\,C\} = \dfrac{P\{C\,|\,\tilde{N}\}P\{\tilde{N}\}}{P\{C\,|\,\tilde{N}\}P\{\tilde{N}\} + P\{C\,|\,N\}P\{N\}}$

   $= \dfrac{(.80)(.10)}{(.80)(.10) + (.20)(.90)} = .31$

2. $P\{N\,|\,\tilde{C}\} = \dfrac{P\{\tilde{C}\,|\,N\}P\{N\}}{P\{\tilde{C}\,|\,N\}P\{N\} + P\{\tilde{C}\,|\,\tilde{N}\}P\{\tilde{N}\}}$

   $= \dfrac{(.80)(.90)}{(.80)(.90) + (.20)(.10)} = .97$

(*Note.* $P\{\tilde{C}\,|\,N\} = 1 - P\{C\,|\,N\}$ and $P\{\tilde{C}\,|\,\tilde{N}\} = 1 - P\{C\,|\,\tilde{N}\}$.)

3. $P\{\tilde{N}C\} + P\{N\tilde{C}\} = P\{\tilde{N}\,|\,C\}P\{C\} + P\{N\,|\,\tilde{C}\}P\{\tilde{C}\}$

   $= (.31)(.26) + (.97)(.74)$

   $= .80$

These statistics can be readily interpreted. If a person receives a C-score, the probability that he is psychotic is .31 and the probability that he is not psychotic is about .69. Such a person ought to be examined by other instruments and interviews. On the other hand, there is little chance ($P = .03$) that a person receiving a $\tilde{C}$-score is psychotic. The use of the screening device over a long period would lead to 80% correct judgments and 20% incorrect judgments, and these incorrect judgments would consist largely of classifying normal subjects as psychotic. A small percentage of psychotics would be classified as normals.

It is not necessary to use the formula for doing the calculations. If you can construct the two-way table correctly, the probabilities can be calculated without formulas by taking appropriate numerators and denominators. In the example, a frequency table can be constructed taking 100 subjects to be a total figure. The correct table is presented below.

	C	$\tilde{C}$	$C + \tilde{C}$
N	18	72	90
$\tilde{N}$	8	2	10
$N + \tilde{N}$	26	74	100

It is enlightening to study the tabled values in relation to the terms of Bayes' equation. For example, $P\{\tilde{N}\,|\,C\}$ can be calculated directly. The C

column is "given" so we focus attention on it. The entry for $\tilde{N}$ is 8, so the answer is $\frac{8}{26} = .31$. In the expanded formula, the denominator is merely the expansion of $P\{C\}$ (see the development of the equation) and $P\{C\}$ is $\frac{26}{100} = .26$. The numerator is $P\{C\tilde{N}\}$ which is $\frac{8}{100} = .08$. So we could also get the answer by the relationship $P\{\tilde{N} \mid C\} = P\{\tilde{N}C\}/P\{C\}$, which also yields .08/.26, or .31. The problem of using the table approach is in getting the correct figures into the table. Recall that each numeral must come from the four basic bits of information $P\{\tilde{N}\} = .10$, $P\{C\} = .26$, $P\{C \mid \tilde{N}\} = .80$, and $P\{C \mid N\} = .20$. See if you can reconstruct the table from these four values.

PROBLEM 17.9

Suppose that in a particular graduate school no screening device has been used and that, in the past, 60% of all students eventually earned their master's degree and the other 40% dropped out for academic purposes. Suppose the school decides to use the Graduate Record Examination (GRE) and will set the admission cut-off as 1000 points, a score attained by 50% of all persons taking the test in a national sample. Previous research indicated that 70% of all graduates scored above 1000 and that only 20% of all students who eventually dropped out scored above 1000. Let G mean "graduate," and A mean "scored above 1000 points." Find the following probabilities by formulas and by table.

a. $P\{G\}$
b. $P\{\tilde{G}\}$
c. $P\{A\}$
d. $P\{\tilde{A}\}$
e. $P\{A \mid G\}$
f. $P\{A \mid \tilde{G}\}$

g. $P\{G \mid A\}$
h. $P\{\tilde{G} \mid A\}$
i. $P\{G \mid \tilde{A}\}$
j. $P\{\tilde{G} \mid \tilde{A}\}$
k. $P\{GA\} + P\{\tilde{G}\tilde{A}\}$

Sampling

The researcher is generally confronted with the problem of studying a very large number of things. Generally, the number is so large that it is impossible, too expensive, or too time consuming to study each of the things. So the researcher resorts to studying a subset of the entire set. This subset, or *sample*, is used to make inferences about the entire set, or *population*. Fortunately, by the proper choice of the *sample*, valid inferences can be made about the *population*.

Consider a school that has 100 students. A researcher wishes to interview the students, and the interview will take one hour. If the interviewer has only 10 hours, then he can talk with only 10 students and must infer that the opinions of these 10 adequately reflect the opinions of the 100. His choice of the 10 students is critical. There are many ways to choose them, and these sampling methods are usually discussed in elementary statistics classes.

Consider the magnitude of the problem. There are, in fact, $\binom{100}{10}$ *different* samples possible. This is $(100!)/(90!)(10!)$. You don't need to multiply this out to tell that there is a very large number of possible samples of size 10 in a population of size 100. You might try these. How many samples of size 5 are there in a population of size 10? How many samples of size 3 are there in a population of size 15? (*Answers.* 252 and 455.)

Of the large number of possible samples, some have characteristics typical of the entire population and some are atypical. The researcher wishes to avoid the atypical samples—such as getting just the very bright children *or* the very dull children when he wishes to study typical children. But getting representativeness is very difficult because of the many ways bias can enter unintentionally into research.

Several techniques have been developed for determining the elements of the sample and these will surely be discussed in your statistics classes. The most important of these techniques is "*simple random sampling.*" The adjective "simple" distinguishes this basic procedure from several others that are derivations of simple random sampling.

Any card game provides a good example of random sampling. "Shuffling" is supposed to *randomize* (or to make random) the deck of cards. "Dealing" is the actual act of sampling. The poker player gets a "random sample" of five cards from the "population" of 52 cards.

Random sampling can be accomplished in two ways. The two ways are similar enough to yield no practical differences in research practice, but subtle differences do exist between them.

The first way we can call *random sampling with replacement*. It is exactly like a lottery in which ticket stubs are well mixed, a number is drawn and recorded, and the *stub is returned to the box* before a second winner is chosen. Thus, it is possible, although unlikely, that a ticket will win two or more prizes. Each draw has *no effect* on subsequent draws.

The second technique is similar. This method is exactly like a lottery in which each winning ticket is set aside and cannot win again. This method can be called *random sampling without replacement*. In this technique, every possible sample has the same probability of inclusion as every other sample.

Actually, a lottery technique is often used. Names or identification numbers are assigned to elements of the population and the sample is identified by lottery. Many other ways of getting random samples and "nearly" random samples are discussed in elementary applied statistics texts.

Distributions

PROBABILITY DISTRIBUTIONS

Probability distributions show the relative frequency estimates of probability of alternative outcomes to an experiment. Distributions can be empirical or theoretical, depending, of course, on whether they are constructed from data or generated from theory, respectively.

Let's take as the first example, the simplest of the theoretical probability distributions (or *probability density* distributions or functions). The tossing of a coin has two outcomes, H or T. The entire probability distribution is presented in the table below.

Outcome	H	T
Probability	$\frac{1}{2}$	$\frac{1}{2}$

Probability distributions are almost always based on mutually exclusive outcomes. The total of all values is unity (in the example, we have $\frac{1}{2} + \frac{1}{2} = 1$) since all possible outcomes are listed.

Let's take another theoretical example, the throw of a single die. The distribution is tabled below.

Outcome	1	2	3	4	5	6
Probability	$\frac{1}{6}$	$\frac{1}{6}$	$\frac{1}{6}$	$\frac{1}{6}$	$\frac{1}{6}$	$\frac{1}{6}$

Check the sum of the entries.

Let's use the last example to see how distributions can be used. By reading the table and adding proper values, the probabilities of combinations of events can be readily calculated. What is the probability of getting a six? The value "$\frac{1}{6}$" can be read directly from the table. What is the probability of getting an even number? The answer can be obtained by adding the probabilities of a two, four, and six to yield $\frac{1}{2}$.

A question that the researcher quite commonly asks has the form, "What is the probability of getting a value equal to or larger than ...?" In this example, we might ask, "What is the probability of getting a value of three or higher?" By adding, the correct answer is $\frac{2}{3}$.

PROBLEM 17.10

Suppose an intelligence test has been given to 1000 subjects to generate an empricial probability distribution of test scores. It would take an extremely large table to show all outcomes (or scores). It is standard practice to group values into intervals, and this is done in the following table.

Probability	.04	.11	.32	.37	.14	.02
Score interval	0–70	71–85	86–100	101–115	116–130	131+

Use these empirical probabilities to estimate the probability of a child scoring:

a. Less than 70.
b. At least 131.
c. At least 101.
d. More than 85.
e. At least 86, but not more than 115.

CUMULATIVE PROBABILITY DISTRIBUTIONS

Another tool of the statistician is the *cumulative probability distribution* (or *cumulative density function*), which is a close relative of the probability distribution. This distribution is a listing of the probability of a particular value *or all lower values*. For example, the die problem gives the cumulative probability distribution that is in the table below.

Cumulative probability	$\frac{1}{6}$	$\frac{1}{3}$	$\frac{1}{2}$	$\frac{2}{3}$	$\frac{5}{6}$	1
Outcome	1	2	3	4	5	6

This table is used to find directly the answers to questions like, "What is the probability of getting no more than a four?" (Answer—$\frac{2}{3}$.) Or, "What is the probability of exceeding a four?" The answer is obtained less directly. It is one minus the probability of getting four or less; that is, $1 - \frac{2}{3}$, or $\frac{1}{3}$. Note that the last value in the table is unity, showing that the roll of the die is certain to yield one of the tabled values.

The intelligence test data provide another example. Be sure that you can construct the table below from the previously presented probabilities.

Cumulative probability	.04	.15	.47	.84	.98	1.00
Score interval	0–70	71–85	86–100	101–115	116–130	131+

SOME COMMON DISTRIBUTIONS

Rectangular Distributions. This distribution results if each possible outcome has the same probability—such as the six outcomes in the toss of a single die. The probability of a number occurring on a spin of a roulette wheel is the same for all numbers if the wheel is honest.

The Binomial Distribution. This distribution is appropriate in the case that an experiment is a combination of events having *two* mutually exclusive outcomes (*two* from the *bi* in binomial). Examples of binomial events are coin tosses, sex determination, success-failure, win-lose, etc.

Coin tossing provides a good example of this important set of distributions. Consider an experiment in which six coins are tossed. We wish to know the probabilities of getting exactly no heads, one head, two heads, etc., up to six heads. The solution is based on the discussion of combinational problems in the preceding chapter. Pascal's Triangle offers a quick solution. Take the sixth row of Pascal's Triangle and add up all the entries. The correct probabilities are obtained by dividing each entry by the sum of all entries. The procedure is outlined below.

Number of heads	0	1	2	3	4	5	6
Sixth row of Pascal's Triangle	1	6	15	20	15	6	1
(Sum is 64) Probability	$\frac{1}{64}$	$\frac{6}{64}$	$\frac{15}{64}$	$\frac{20}{64}$	$\frac{15}{64}$	$\frac{6}{64}$	$\frac{1}{64}$

Check to be sure that the probabilities sum to one.

It will help to relate this discussion to the previous discussion of the binomial expansion. The probabilities in the table are the terms in the binomial expansion of $(a + b)^6$. Pascal's Triangle gives the coefficients $\left(\text{or these can be calculated by the combinational operations} \binom{6}{0}, \binom{6}{1}, \binom{6}{2}, \right.$ etc.$\left.\right)$. Now, if you set $a = \frac{1}{2}$, which is the probability of a head, and $b = \frac{1}{2}$, which is the probability of a tail, then each of the values $\binom{6}{0}, \binom{6}{1}$, etc., is multiplied by the *same* number, namely $(\frac{1}{2})^6 = \frac{1}{64}$, to yield the desired probabilities.

PROBLEM 17.11

Suppose a rat is running a simple maze. The rat can make three choices, only one of which is correct. If the rat runs the maze four times, what is the probability distribution of the number of correct choices—assuming that the rat runs at random? Return to the previous chapter for the expansion of $(a + b)^4$. Set $a = \frac{2}{3}$ (the probability of an incorrect choice) and $b = \frac{1}{3}$ (the probability of a correct choice). Calculate the probabilities.

Suppose a rat is put through the series of mazes and makes all four choices correctly. The experimenter might then conclude that the rat "knows" the right choice, because the probability of getting four correct responses by *chance* is $\frac{1}{81}$. Since this probability is relatively small, it is plausible to believe that the rat was acting systematically, and not randomly.

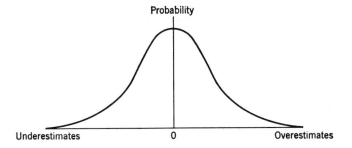

Figure 17.2   The normal distribution of errors.

The Normal Distribution. Of all theoretical distributions, the most famous is the *normal distribution*. This distribution is named because of the large number of times it occurs in nature and the wide use of it in applied statistics. You will spend a large amount of time in your statistics classes learning the use and characteristics of the normal distribution (or *normal curve*, as it is sometimes called). Since the statistics class will emphasize this topic, there is no need to discuss it at length here.

A most common application of the normal curve is in the distribution of errors of observation. This example can serve as a device for discussing the general characteristics of the distribution. The observations can be of anything. We can be estimating length, measuring intelligence, or predicting job success. Let's assume that we are discussing chance, or random error, and have no systematic bias in our observing technique. Hopefully, a large number of observations will have very little error—slight overestimates and slight underestimates. Also, we hope that few observations will yield large errors, either over- or underestimates. In general, we expect that the probability of an error decreases as the size of the error increases. The normal distribution is a good model for such error distributions. It allows for a "pileup" of values near the average error of zero and a decreasing probability for large errors.

The general shape of the normal distribution is a bell (it is frequently called "the bell curve"). The picture of a bell would result if the probability values for each outcome were graphed. The result would be similar to Figure 17.2.

In addition to the distribution of error, the normal curve has many other important uses. Two of these are the making of statistical tests of hypotheses and the interpretation of psychological tests. For example, it is common to assume that intelligence and school achievement have normal distributions, and so intelligence tests and achievement tests are usually interpreted and constructed with the aid of the normal probability distribution.

Problems for Review

1. Two cards are drawn from a well-shuffled deck. Calculate the probability that:

  a. Both are spades.
  b. One is a spade and one is not a spade.
  c. The first card is either a spade or a face card (J, Q, or K).
  d. The first card is a spade and the second card is a face card.
  e. The first card is either an ace, deuce, or a spade.

2. In a hypothetical university there are 1000 students, 400 of whom scored high on the verbal CEEB Exam and 600 of whom scored high on the Quantitative Exam. 125 students were high on both exams. Let S be the set of all 1000 students, V be the high-verbal students, and Q be the high-quantitative students. See Figure 17.3.

  Calculate the requested probabilities. Be sure you can translate the set symbolism into meaningful English.

  a. $P\{\text{Bob} \in V\} = ?$ e. $P\{\text{Bob} \in V \mid \text{Bob} \in Q\} = ?$
  b. $P\{\text{Bob} \in Q\} = ?$ f. $P\{\text{Bob} \in Q \mid \text{Bob} \in V\} = ?$
  c. $P\{\text{Bob} \in V \cap Q\} = ?$ g. $P\{\text{Bob} \notin Q\} = ?$
  d. $P\{\text{Bob} \in V \cup Q\} = ?$ h. $P\{\text{Bob} \notin Q \mid \text{Bob} \notin V\} = ?$

		College Average (C)			
		1.0	2.0	3.0	4.0
High school	4.0	50	100	200	300
average (H)	3.0	100	200	300	150
	2.0	200	400	200	100
	1.0	500	100	50	50
	Sum	850	800	750	600

3. The above table is a *joint-frequency distribution table*. It is often called an "experience chart" or "expectancy table." It shows the number of persons (hypothetical) out of 3000 who earn various grades in high school and college. For example, the first cell means that 50 students who were 4.0 students in high school received 1.0 averages in college. This type of table is

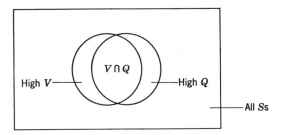

Figure 17.3

used often in academic prediction studies. Let H stand for "high school average" and C stand for "college average." Calculate the following probabilities. Be sure you know what each means in terms of the problem.

a. $P\{C = 4.0\}$

b. $P\{C = 3.0 \text{ or } 4.0\}$

c. $P\{C = 4.0 \mid H = 4.0\}$

d. $P\{C = 3.0 \text{ or } 4.0 \mid H = 4.0\}$

e. $P\{C = 3.0 \text{ or } 4.0 \mid H = 1.0 \text{ or } 2.0\}$

f. $P\{C = 4.0 \mid H = 1.0\}$

g. $P\{C = 4.0 \text{ and } H = 4.0\}$

h. $P\{C = 2.0 \text{ and } H = 2.0\}$

i. $P\{C = H\}$

4. Suppose there exists a hypothetical state that has four counties. The population of each of the counties is as follows.

County	Population
1	200,000
2	80,000
3	100,000
4	20,000

a. If a simple random sample of 8000 (2% of 400,000) is drawn from this state's population, what is the probability of including a particular person, Mr. Jones?

b. If Mr. Jones resides in county 4, and the *entire* 2% is chosen from any *one* county which is chosen at random, what is the probability that Mr. Jones will be chosen?

c. Suppose the selection of the one county is based on the size of the population of the counties, i.e., county 1 is given 10 times the probability of being chosen as county 4 since it is 10 times as large. What is the probability of Mr. Jones being chosen with this sampling method?

d. Suppose Mr. Smith lives in county 1. What is the probability of his being chosen under each above condition (*a*, *b*, and *c*)?

e. Based on these answers, evaluate the three sampling methods implied by Problems *a*, *b*, and *c*.

Achievement quartile		I	II	III	IV	V
	IV	0	0	35	45	20
	III	0	5	65	25	5
	II	5	25	65	5	0
	I	20	45	35	0	0

Socioeconomic Status

5. This expectancy table shows hypothetical frequencies in which persons are cross classified on socioeconomic class and achievement. Assume that the table is a valid and reliable reflection of the relationship between socioeconomic status and school achievement. Suppose a child is chosen at random.

a. What is the probability that he is in the top achievement category (marginal probability)?

b. What is the probability that he is in the lowest SE category (marginal probability)?

c. What is the probability that he is *both* in the middle SE group (III) *and* in the top achievement group (IV) (joint probability)?

d. What is the probability that he is *both* in the top SE category *and* the *highest* achievement category (joint probability)?

e. What is the probability that he is in *both* the *lowest* SE category *and* the *highest* achievement category (joint probability)?

f. What is the probability that he is in *either* of the lowest two SE categories (marginal probability)?

g. What are the chances that a middle-class child (category III) will be in the lowest achievement group?

h. Suppose you know that a child has an achievement score in the group III range. What is the probability that he is in the highest SE group?

i. What is the *marginal probability distribution* of socio-economic status? (Be sure all values sum to unity.)

j. What is the *conditional probability distribution* of achievement for the lowest socio-economic group? (Be sure all values sum to unity.)

k. What is the cumulative probability distribution for socioeconomic status?

6. Suppose 10 children are chosen at random.

a. What is the probability that *all* of these children are in the upper half of the general population with respect to intelligence (that is, their IQ is above 100)?

b. Suppose that you measure their intelligence and find that, in fact, *all* do have intelligence scores above 100—what would you conclude?

c. What is the probability that three or fewer persons chosen at random will have IQ's higher than 100?

7. Suppose you are playing bridge. You know only that you and your partner have nine spades between you. What is the probability that the remaining spades are split:

a. 4 — 0?

b. 2 — 2?

c. 3 — 1?

d. 3 — 1 with the player on your left holding three? *Hint.* First calculate
$$\binom{4}{0}, \binom{4}{1}, \binom{4}{2}, \binom{4}{3}, \text{ and } \binom{4}{4}.$$

8. A "flush" is a poker hand consisting of five cards of the same suit. What is the probability of being dealt (a) a spade flush? (b) a flush of any suit? (c) a royal flush (10, J, Q, K, A of any one suit)?

9. Suppose there exists a bad teacher who has taught nothing. He gives an eight question true-false test to his class and they answer by guessing. What is the probability distribution of the number of correct answers?

Part Five

MISCELLANEOUS SKILLS

THIS part consists of several unrelated topics. One chapter broadens comments made earlier relative to irrational numbers. The other chapters deal with specific skills useful in statistical studies.

Two "skill chapters" are substantive in that important concepts, as well as skills, are involved. The first of these, on graphs and graphing, will have numerous applications in beginning and advanced statistics courses. The second, on logarithms, is of less importance in terms of the frequency of use; however, logarithms are occasionally used in early statistics training, so it is appropriate to have this chapter for reference.

The other two "skill chapters" concern table reading and computational accuracy. These chapters deal with skills that are used daily in statistical training and applications.

18

Irrational Numbers

Introduction

IRRATIONAL numbers have been mentioned from time to time because of their importance. It is appropriate to give additional consideration to the difference between rational and irrational numbers. Special attention will be given to some particular irrational numbers called "π" and "e," which are used commonly in statistical problems.

Irrational numbers have been defined previously as numbers that cannot be expressed exactly as numerals in ratio (common fraction) form or as decimal fractions. They can be approximated in decimal notation. Why should we use them then? We use them because they exist and are as real as rational numbers. Rational numbers and irrational numbers are both subsets of the real number system. The problems of irrational numbers arise not from the numbers themselves, but in the limitations of decimal notation. If we used another base to our number system, then some of the numbers we now call rational would not be rational and some we call irrational would be rational.

Here is a very common example. Suppose we have a square with a side one unit long. How long is the diagonal of the square? It cannot be stated exactly in decimal form. One can use a ruler and measure it fairly accurately, but not exactly, if the ruler is calibrated in any usual form. The diagonal is precisely $\sqrt{2}$ units long, where $\sqrt{2}$ is irrational. To say "it is $\sqrt{2}$" is perfectly precise. There is no better way to answer the problem in terms of mathematical precision. A decimal *approximation* may be more useful and can be determined with a great deal of accuracy.

If we get an irrational number with such a simple physical problem, how many other irrational numbers are there? There is an infinite number of them. In fact, most square roots are irrational. The number $\sqrt{x}$ will *always* be irrational *if x is not negative and is not a perfect square.*

221

Let's look at $\sqrt{3}$. If you are asked, "What is the square root of 3?" you probably would think that the question meant, "What decimal fraction is another name for $\sqrt{3}$?" You would set out to determine this value by the method presented in an earlier chapter. Let's look at some approximations. It is less than 2 because $2^2 = 4$. It is greater than 1 because $1^2 = 1$. Let's split the difference between 1 and 2 to get 1.5. But $\sqrt{3} > 1.5$, because $1.5^2 = 2.25$. So, $1.5 < \sqrt{3} < 2$. Let's try 1.75. We get $1.75^2 = 3.063$, so $1.5 < \sqrt{3} < 1.75$. But 1.75 is obviously close, so let's try 1.73 and 1.74. We get $1.73 < \sqrt{3} < 1.74$. We can continue this process indefinitely, and narrow the interval until it is as small as we wish, but we will never exactly pinpoint $\sqrt{3}$ in decimal notation.

Let's digress for a moment and discuss rational numbers for awhile. Insight into rationals will provide additional insight into irrationals.

Rational Numbers

INTRODUCTION

Every *rational* number can be expressed precisely as a ratio of two integers. Here are some examples:

$$\tfrac{1}{2}, \tfrac{3}{2}, \tfrac{7}{1}, \tfrac{2}{3}, \tfrac{1}{6}, \text{ etc.}$$

Any common fraction is a rational number. This is, in fact, the defining property of rational numbers.

If a number is rational, then it can be written in decimal notation by carrying out the division indicated by the fraction (e.g., $\tfrac{1}{2} = 1 \div 2 = .5$). Two cases arise. The division may come out even (e.g., $\tfrac{3}{2} = 3 \div 2 = 1.5$). This is called a *terminating* decimal. The division may not come out even. For example, $\tfrac{1}{3}$ is $1 \div 3$, which gives the form ".33333...," where the three dots mean that the "3's" continue indefinitely. A bar above decimals is often used to indicate repeating decimals. We could write "$\tfrac{1}{3}$" as ".$\overline{3}$" or ".$3\overline{3}$." Here are more examples:

$$.\overline{85} = .8585858...$$
$$2.\overline{123} = 2.123123123...$$
$$4.10\overline{12} = 4.10121212...$$
$$3.751\overline{2} = 3.75122222...$$

A general principle of rational numbers is that each can be expressed not only as a ratio, but also as a repeating decimal *or* as an exact (terminating) decimal with no approximation error. Irrational numbers can be written in *neither* way.

The conversion of ratios into decimals is obviously merely the division process. The conversion of decimals into ratios is slightly more complex and will be discussed in the next few sections.

CONVERSION OF DECIMALS INTO RATIOS

Terminating Decimals. Terminating decimals are converted into ratios by choosing the proper denominator. This is quite simple:

$$1.1 = \frac{11}{10} \qquad\qquad .0124 = \frac{124}{10,000}$$

$$3.45 = \frac{345}{100} \qquad\qquad .51 = \frac{51}{100}$$

$$7.321 = \frac{7321}{1000} \qquad\qquad .1032 = \frac{1032}{10,000}$$

Many of these can be simplified to some degree by dividing numerator and denominator by constants, but this step is unnecessary.

Repeating Decimals. Repeating decimals are a little more difficult to convert into ratios. Let's do some examples. Each is solved by constructing two equations and solving for one unknown. Let K be the repeating decimal that we are trying to write in ratio form:

$$K = .6\overline{6}$$

$$
\begin{array}{l}
10K = 6.66\overline{6} \\
\underline{K = .66\overline{6}} \\
9K = 6 \\
K = \tfrac{6}{9} \text{ or } \tfrac{2}{3}
\end{array}
$$

$$K = 1.3\overline{3}$$

$$
\begin{array}{l}
10K = 13.3\overline{3} \\
\underline{K = 1.3\overline{3}} \\
9K = 12 \\
K = \tfrac{12}{9} = 1\tfrac{1}{3}
\end{array}
$$

$$K = .24\overline{2424}$$

$$
\begin{array}{l}
100K = 24.2\overline{424} \\
\underline{K = .24\overline{24}} \\
99K = 24 \\
K = \tfrac{24}{99}
\end{array}
$$

$$K = .1762\overline{1762}$$

$$
\begin{array}{l}
10,000K = 1762.\overline{1762} \\
\underline{K = .\overline{1762}} \\
9999K = 1762 \\
K = \tfrac{1762}{9999}
\end{array}
$$

Can you see the pattern? The equations are chosen so that the repeating portions of the numbers subtract out, leaving only integers. This is a process which obviously cannot be done with irrationals since there is no repeating portion to subtract out.

The Numeral "π"

One of the irrational numbers that arises quite often in statistics is called "pi" (π). The number π arises in the study of circles. For *every* circle, the ratio of the circumference length to the diameter length is one number. This ratio, common to all circles, is called "π" and is an irrational number.

The number π is irrational, so it cannot be written precisely as a ratio, a terminating decimal, or a repeating decimal. However, it can be estimated in decimal notation as accurately as we wish. Common estimates of π are 3.14, 3.142, and 3.1416. The ratio $\frac{22}{7}$ is also used as an estimate of π. But keep in mind that all of these numbers do not equal π exactly.

The number π arises quite often in mathematical problems involving geometric concepts. It is especially useful in problems dealing with angle measurement. It is most commonly seen in statistics as a constant in the equation for the bell curve (normal probability curve).

The Numeral "e"

The letter "e" is used as the symbol referring to another important irrational number. The number e is also a constant in the equation for the normal probability curve. It appears in other statistical formulas, too, but its most important value is as the base for "natural logarithms," which are discussed in the next chapter. Its use in logarithms is common in statistical problems. The number e also has wide application in calculus.

The number e can be defined as $\left(1 + \dfrac{1}{n}\right)^n$, where n is a very large number. The proper statement is that e equals the limit of $\left(1 + \dfrac{1}{n}\right)^n$ as n becomes very large. Let's look at this function as n does become large. (See top of page 225.) As n increases, the estimate of e continues to increase. However, it never surpasses a particular constant (it reaches a limit). This limit is e. Estimates of e that are often used are 2.72, 2.718, and 2.7183. Correct to eight decimal places, e is 2.71828182.

n	$\left(1 + \dfrac{1}{n}\right)^n$
1	$2.000\overline{0}$
2	$2.250\overline{0}$
3	$2.\overline{370}$
4	$2.4414\cdots$
10	$2.594\cdots$
1000	$2.717\cdots$

Problems for Review

Determine if the solution sets to these equations contain rational or irrational numbers. If they are rational, express them in both ratio and decimal notation.

1. $x^2 + 1 = 4$

2. $x^2 = 2.56$

3. $36x^2 = 1$

4. $8x^2 = 2$

5. $(x + 1)^2 = 0$

6. $9y + 20 = 45$

7. $3y^2 - 1 = 4$

8. $x - 1 = \left(\dfrac{1}{10} + \dfrac{1}{100} + \dfrac{1}{1000} + \dfrac{1}{10,000} + \cdots\right)$

9. $x^2 - 7x + 12 = 0$

10. $N = (1.101001000100001\cdots)$

19

Graphing

Introduction

GRAPHING is a major statistical tool that is used in almost every statistical procedure from simple data description to complex statistical analyses. Some special graphing procedures are generally taught as a part of a statistics course. However, it is necessary to know some fundamentals of graphing.

Let's consider the fundamental mathematical concept of graphing before we talk specifically about Cartesian coordinate systems. Think of two ships standing still on the ocean. They are only a few miles apart. How might you describe their relative positions? You might say that "ship A is 10 miles northeast of ship B." You have chosen ship A as a reference point (arbitrarily) and described ship B's location relative to the location of ship A. You could have said similarly, "Ship B is 10 miles southwest of ship A." Here, you have chosen ship B as an arbitrary reference point. In either case, the relative position of the ships was described by two pieces of information (i.e., two variables, or two dimensions)—*distance* apart and *direction* from one to the other.

A second way to describe the location of the two ships is by giving the longitude and latitude locations of each. The reference point is still arbitrary—it just happens to be a fairly well used one—the point at which the equator crosses the line of zero longitude. Again, two numbers are required to locate each ship.

If our example had been to describe two points on the earth's surface, we might choose to use a three-dimensional coordinate system corresponding to latitude, longitude, and altitude.

Usually, however, a statistician is not interested in describing points on the earth's surface, nor in our stellar space, unless he is working on a geographical or astronomical problem. It is still customary, however, to talk about "locating a point in space," since the statistical problem of graphing is directly comparable to the examples quoted—we want to choose a relatively arbitrary reference point and describe the relationship between quantities in terms of the reference point and some sort of coordinate system.

Specifically, in two dimensions, these "quantities" are ordered number pairs (or vectors of order two). We wish to construct a drawing showing the relationship between various vectors.

The graphic description of data or numeric relationships does involve considerable subjectivity. The choice of the reference point and the scales are two decisions that must be made. Another is the type of graph. The most widely used type of graph is the rectangular coordinate system invented by Rène Descartes (and thus called the "Cartesian coordinate system"). Due to its wide use, it will be the major topic of this chapter. Other coordinate systems will be mentioned, but you should concentrate on learning the effective use of Cartesian graphing.

This chapter will also be limited largely to the construction of graphs involving two variables—that is, two-dimensional graphs. We have already used one-dimensional graphs in the section on negative numbers (the number line). Graphs with more than two dimensions will be mentioned, but the focus of the chapter will be on two variable problems.

Statistical graphs are used, among other things, to describe data, to picture statistical relationships, and to illustrate theoretical models. Graphs can be crude sketches used by a researcher to help himself understand his problem. They can be very well drawn and carefully scaled and published in documents or journals that are widely distributed. Even published graphs can range from sketches that indicate general trends to highly accurate charts intended for use in problems where accuracy is important.

For pedagogic purposes, however, fairly rough sketches will generally be sufficient.

Cartesian Coordinate Systems

A two-dimensional Cartesian coordinate system is a reference system for locating ordered number pairs on a plane by using two reference axes that meet at a right angle to each other. Such a system is usually sketched with one axis vertical and the other axis horizontal, as in Figure 19.1. The point of intersection is called the "origin of the system." The graph is labeled by starting at the origin and assigning numerals to the units of the graph. Units can have intervals of one, two, five, 500, or any size convenient to the problem of interest. Figure 19.2 shows the horizontal axis labeled in units of size one. Both axes are labeled in Figure 19.3. Notice that the origin is the zero point for both the horizontal and the vertical axes.

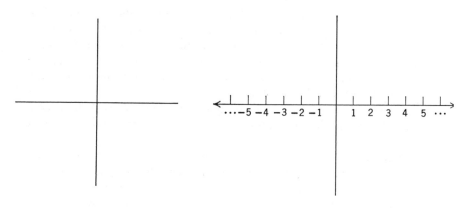

Figure 19.1   *Figure 19.2*

The locating of points on the graph is called "plotting." Points are described as ordered number pairs (vectors of two elements). It is customary to consider the first of the two numbers in the ordered pair as referring to the horizontal axis. The second number, then, refers to the vertical axis. Consider, for example, the ordered pair (4, 6). The location of the point on the graph can be considered similar to the problem of locating an address on a city map. The pair (4, 6) says, in effect, "go from the origin 4 units on the first (horizontal) axis, then go 6 units on the second (vertical) axis." Figure 19.4 shows a graph on which (4, 6) is plotted. It also shows (4, −6), (−4, −6), and (−4, 6). Notice that a minus sign means "go left" or "go down," depending on whether the negative number is first or second in the pair. Study Figure 19.4 and be sure you can tell why the points are plotted as they are.

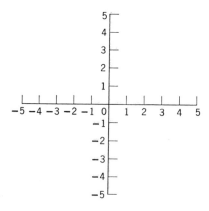

Figure 19.3

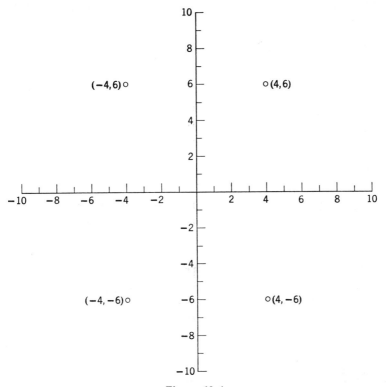

Figure 19.4

Here are some more fundamental principles. The first number in the number pair is called the "*abscissa* of the point." The second number is called the "*ordinate*." These words will be used often, so you should learn to use them. The Cartesian graph is divided into four quadrants by the pair of reference axes. The first quadrant (I) consists of all points that have positive abscissas and positive ordinates. Find the quadrant in Figure 19.5. Figure 19.5 also shows quadrants II, III, and IV. Notice the signs of the abscissas and ordinates for the four quadrants. Let's summarize these.

Quadrant	Sign of Abscissa	Sign of Ordinate
I	+	+
II	−	+
III	−	−
IV	+	−

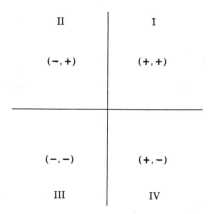

Figure 19.5

PROBLEM 19.1

Now you should stop and practice. Here are some number pairs. Construct a graph and locate each number pair on the graph.

1. (0, 0)
2. (3, 3)
3. (−5, −3)
4. (3, 6)
5. (1, −6)
6. (−7, 6)
7. (10, 10)

8. (−4, 3)
9. (−8, −9)
10. (5, −2)
11. (0, 10)
12. (10, 0)
13. (−10, 0)
14. (0, −10)

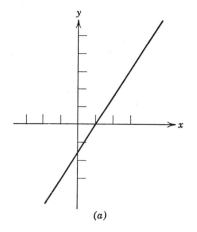

(a)

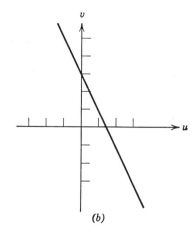

(b)

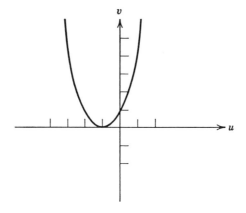

Linear Functions

INTRODUCTION

Cartesian graphs are convenient for studying functional relationships. Linear functions are of special interest. A function is linear if it can be written in the form

$$y = ax + b$$

where a and b are constants and y and x are variables. The equation can be solved for x, which also will yield a linear function. One gives y as a function of x and the other gives x as a function of y.

The graphs can be constructed in several ways. Let's use an example. $Y = X$ is a linear function. We can use it to define a set of ordered number pairs, but we must be careful to note which of the numbers in the pair correspond to X and which to Y. Here is a set of pairs (X, Y): $(-2, -2)$; $(-1, -1)$; $(0, 0)$; $(1, 1)$; $(2, 2)$; and $(3, 3)$. These points are plotted in Figure 19.6. They are connected with a straight line. The straight line in Figure 19.6 gives *all* points (X, Y) such that $X = Y$, if you extend it indefinitely. For example, it passes through $(-.25, -.25)$ and $(\sqrt{3}, \sqrt{3})$.

Now, it was not necessary actually to plot all of those points because a *straight line is determined by only two points*. Let's look at the function "$Y = 2X$." Let's choose two numbers for X and find the corresponding Y values. Any numbers will do, but zero is usually a convenient one. So one (X, Y) point might be $(0, 0)$. Let $X = 2$ define the second point. If $X = 2$, then $Y = 4$, and a second point is $(2, 4)$. Two points are all that we need. These two points are plotted in Figure 19.7. The points are joined by a straight line. Notice that Figure 19.7 is also the graph of $X = .5Y$.

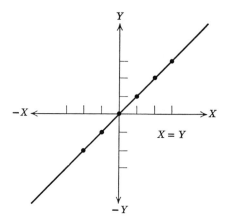

Figure 19.6

Here is another example: $Y = .5X + 1$. What is the second number in these pairs for $Y = .5X + 1$ if the pairs are in the order (X, Y)?

$$(0, ?) \quad \text{and} \quad (2, ?)$$

These points are plotted in Figure 19.8 and the graph has been completed. The function $Y = \frac{1}{3}X - 2$ is plotted in Figure 19.9.

There is another way to graph linear functions. The constants in the functions have a very special graphic meaning. In the general equation

$$Y = aX + b$$

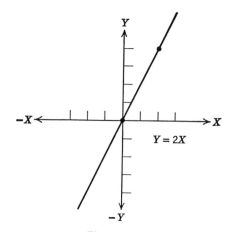

Figure 19.7

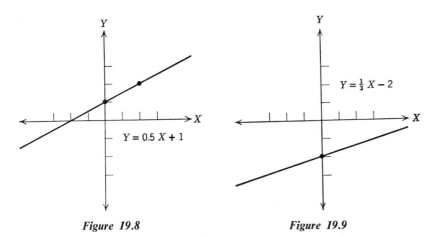

Figure 19.8   **Figure 19.9**

a is the slope of the plotted line and *b* is the point at which the line crosses vertical axis (i.e., the point at which $X = 0$). Check this claim out with the preceding four examples. The word "slope" means, literally, the degree to which the line slopes. It is the number of ordinate units that is increased with each abscissa unit.

In Figure 19.6, the equation is $X = Y$. The constant *b* is zero, so the graph passes through $(0, 0)$. The constant *a* is one, so the graph increases one *Y*-unit for each one *X*-unit (the slope is one).

In Figure 19.7, the equation is $Y = 2X$. Again, $b = 0$ and the line goes through the origin. The slope is two, so the line goes up two units for every one unit that it goes right. And it goes down two units for every unit it goes left.

In Figure 19.8, $b = 1$, so the line passes through $(0, 1)$. In Figure 19.9,

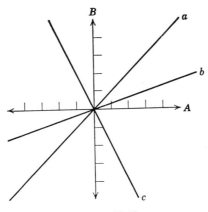

Figure 19.10

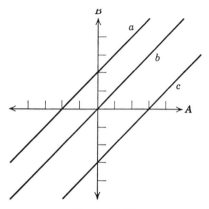

Figure 19.11

$b = -2$, and the line crosses the Y-axis at -2. The slope is $\frac{1}{2}$ in Figure 19.8—see how the line only increases $\frac{1}{2}$ of a Y-unit for every increase of one X-unit. In Figure 19.9 the line only increases $\frac{1}{3}$ of a Y-unit for every X-unit. Its slope is $\frac{1}{3}$. The technique of constructing graphs in this manner is called the "slope-intercept method."

Figures 19.10 and 19.11 demonstrate these principles fairly clearly. Figure 19.10 gives a family of lines, each of which passes through the origin—that is, each has an intercept of zero, but each has a different slope. Figure 19.11 shows a family of lines with the same slope but differing with respect to the intercept. Figure 19.11 demonstrates that lines with the same slope are parallel lines.

PROBLEMS

19.2. In Figures 19.10 and 19.11, determine an equation for each line by examination of the lines.

19.3. Draw graphs of these functions.

a. $S = R + 1$

b. $R = S + 1$

c. $B = 2A + 2$

d. $-2A = B - 2$

e. $Y = 2X - 15$

f. $Y = .1X + .3$

Nonlinear Functions

Although linear functions are of major interest to statisticians, many nonlinear functions are also. The graphing of nonlinear functions is a highly developed science which involves analytic geometry and the calculus. However, the applied statistician can get by in most of his graphing by plotting individual points and joining the points with a curved line.

Let's use a quadratic function as an example. Our example will be $a = b^2 - 2$. Let's construct a table of (b, a) pairs (Table 19.1).

Table 19.1

b	0	.5	1	1.5	2	2.5	3
a	-2	-1.75	-1	.25	2	4.25	7
b		$-.5$	-1	-1.5	-2	-2.5	3
a		-1.75	-1	.25	2	4.25	7

We can plot these points and connect them with a curve to get the quadratic curve. This curve appears in Figure 19.12.

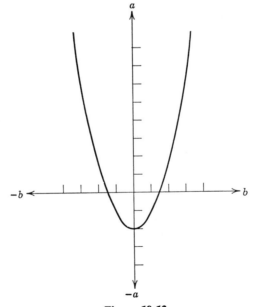

Figure 19.12

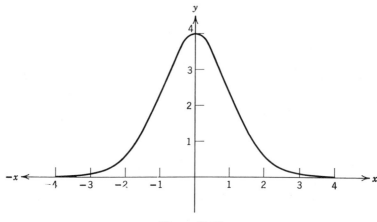

Figure 19.13

Here is another example (Table 19.2) of a nonlinear function. Let X and Y be defined by this set of number pairs.

Table 19.2

X	0	.5	1.0	1.5	2.0	2.5	3.0	3.5
Y	4	3.5	2.4	1.3	.5	.2	.04	.01
Y		−.5	−1.0	−1.5	−2.0	−2.5	−3.0	−3.5
X		3.5	2.4	1.3	.5	.2	.04	.01

This function is plotted in Figure 19.13. This is the well-known normal curve (or bell curve) that plays a predominant role in the study and use of statistics.

Notice how the tails of the curve come down to the zero line, but never quite get to it. This is called an "*asymptote* effect." The curve is said to "approach zero *asymptotically*" or "to approach zero as an *asymptote*." Asymptotes are found fairly commonly in nonlinear curves. The function $y = 1/x$ has both axes as asymptotes, which reflects the reason why division by zero is such a problem. You might wish to sketch out $y = 1/x$ and see how y becomes large as x approaches zero. Use positive and negative values.

There are many nonlinear curves of interest in statistics, but these two examples will give you an idea of how to go about graphing non-linear relationships.

PROBLEM 19.4

Construct a graph from the following pairs of numbers.

b	3.00	1.25	0	−.75	−1.00	−.75	0	1.25	3.00
a	−1.00	−.50	0	.50	1.00	1.50	2.00	2.50	3.00

Solving Equations and Inequalities Graphically

GRAPHIC DETERMINATION OF SOLUTION SETS

Graphs are convenient ways to study equations, equation systems, and inequalities. Solution sets can be demonstrated graphically, whether the set contains no element or an infinite number of elements.

LINEAR EQUATIONS

Let's look first at some linear equations. What is the solution set for the equation $5X - 4 = 0$? The equation can be reformulated into a function of X. $Y = 5X - 4$ is a functional relationship corresponding to $5X - 4 = 0$. We can now rephrase our question this way: "What value or values of X correspond to $Y = 0$?" The graphic solution is to plot $Y = 5X - 4$ and look at the point at which the line crosses the X-axis. Since the X-axis corresponds to $Y = 0$, the intercept of the line with the X-axis is the root of $5X - 4 = 0$. Figure 19.14 shows the line $Y = 5X - 4$ and its intercept with the X-axis $(.8, 0)$. The root is .8.

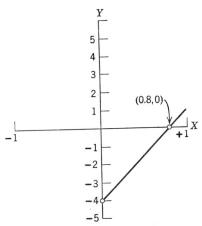

Figure 19.14

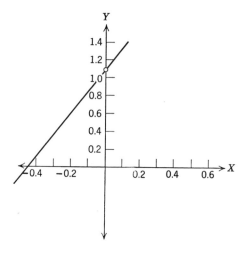

Figure 19.15

Here is another example. Find the root of $2.4X + 1.1 = 0$. First, plot the line $Y = 2.4X + 1.1$ and then find the abscissa (X-value) when the ordinate (Y) is zero. This function is graphed in Figure 19.15. The need for accurate graphing is apparent in this example. The algebraically determined solution set is $X = -1.1/2.4$, or $-.46$. A crude graphic sketch gives a range of possible roots of about $-.4$ to $-.6$. We would probably guess $-.5$ from the graph, which is quite close to $-.46$. More accuracy is most easily obtained by using a larger piece of graph paper and a sharper pencil! A graph with a unit of .01 would give us accuracy to the second decimal place.

NONLINEAR EQUATIONS

Nonlinear equations can also be graphed in order to determine solution sets. In fact, there are occasions when this is the easiest way to do so, especially if a high degree of accuracy is not really mandatory.

For example, what is the solution set to the equation $b^2 - 2 = 0$? The answer is, of course, the irrational number pair $(-\sqrt{2}, \sqrt{2})$. But these values can also be obtained from the graph of $a = b^2 - 2$, which is shown in Figure 19.12. It is quite clear from the intercepts of the curve with the horizontal axis that the solution set consists of two numbers (because there are two intercepts) and these numbers are approximately -1.4 and $+1.4$. Again, if the graph had been drawn with a unit of .01, roots of -1.41 and $+1.41$ probably would have been estimated. (The square root of 2 is about 1.414.)

For a second example, look again at the solution to Problem 19.4 at the end of the last section. No equation was given, but a graph was drawn. Let's ask the question, "Whatever the equation that gave rise to the plotted points, what is its solution set?" The graph clearly gives two roots—0 and 2.

Consider Figure 19.13, the normal curve function. If this function was written as an equation set to zero, what roots could be obtained? The graph shows that there will be *no* roots that will satisfy such an equation.

Many nonlinear curves have no intercepts corresponding to roots. The quadratic curve $X^2 + 1 = 0$ is such an equation. The roots are complex numbers since they involve square roots of -1. You might graph $X^2 + 1 = Y$ and see that the curve does not cross the X-axis.

INEQUALITIES

Inequalities can also be clarified by graphing techniques. The solution sets to inequalities are often infinite sets, so a pictorial representation is often quite helpful. Let's look first at some one-dimensional graphs. Figure 19.16 shows the graph of $X < 5$ and Figure 19.17 shows the graph of $X \leq 5$. The curved line ")" indicates that 5 is not in the solution set. The bracket "]" means that 5 is in the solution set. The solution sets are shown by bold lines. Figure 19.18 shows the graph of $1 < a < 4$.

Now, let's look at two-variable graphs. We'll use shaded areas to show the solution sets. The general method of graphing an inequality, say, $2X - Y \leq 1$, is to graph the boundary line $2X - Y - 1 = 0$, or $Y = 2X - 1$. The direction of the inequality shows on which side of the line the solution set is. The next several figures show, for the two variables a and b, the following solution sets:

Figure 19.19	$a \geq b$
Figure 19.20	$a \leq b$
Figure 19.21	$a \leq 5$
Figure 19.22	$a + b \geq 4$
Figure 19.23	$2a - b \leq 1$

Figure 19.24 shows an example of a type of inequality set that is quite common in statistics. It is

$$1 \leq a \leq 4 \quad \text{and} \quad -1 \leq b \leq 3$$

Figure 19.16

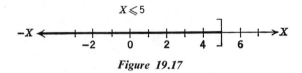

Figure 19.17

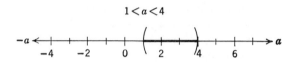

Figure 19.18

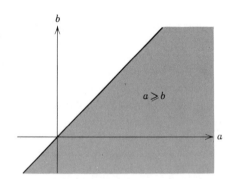

Figure 19.19

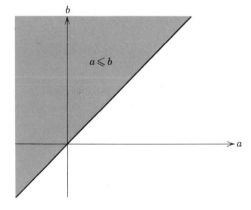

Figure 19.20

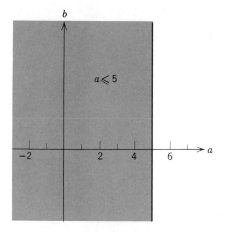

Figure 19.21

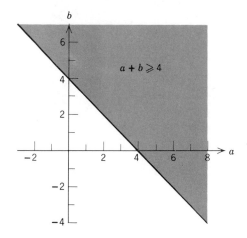

Figure 19.22

SIMULTANEOUS SOLUTIONS TO EQUATIONS

Systems of equations can be solved by graphing. The simultaneous solution to the equations

$$y = 2x - 4$$

$$2y = x + 4$$

is determined graphically by plotting both lines. The point at which the lines intersect is the solution. This is a point that both lines share, thus it is

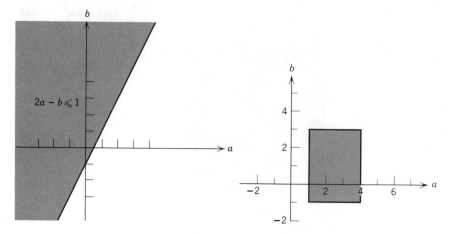

Figure 19.23   *Figure 19.24*

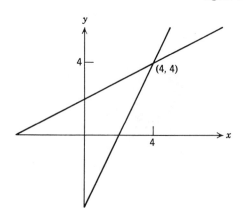

Figure 19.25

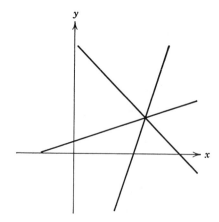

Figure 19.26

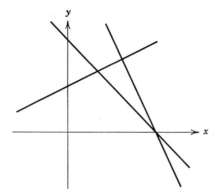

Figure 19.27

a number pair that makes both equations true. The two lines are plotted in Figure 19.25. The point of intersection is (4, 4), so the solution to the two-equation system is $x = 4$ and $y = 4$. Figures 19.26 and 19.27 show three equations in two unknowns. Figure 19.26 shows a set of three equations that have a mutual point in common (i.e., a mutual solution). Figure 19.27, on the other hand, shows three equations that do not have a mutual solution since they do not pass through a single point.

PROBLEMS

Graph each of the following sets of three equations in a single diagram.

19.5. a. $y = 2x + 4$

    b. $y = 2x + 2$

    c. $y = 2x - 1$

19.6. a. $y = x/2 + 1$

    b. $y = 2x + 1$

    c. $y = -x + 1$

19.7. a. $y = x$

    b. $y = 2x - 2$

    c. $y = -2x + 6$

Sketch graphs of these equations and inequalities and show the solution sets.

19.8. $3x - 6 = 0$

19.9. $x^2 - 4 = 0$

19.10. $0 < 2x < 2$

19.11. $-1 < a < 1$ and $a < 2b + 3$

19.12. $x - 2y - 6 < 0$ and $3x + y \geq 4$

Data Plotting

FREQUENCY DISTRIBUTIONS

One of the most common kinds of statistical graphs is a graph showing the frequency of occurrence of some event. Such a graph is called a *"frequency distribution."* (The term also refers to the set of number pairs plotted on the graphs.) The vertical axis of such a graph always shows frequency of occurrence. The horizontal axis is some variable being studied. Such a graph might show how many persons earn certain IQ scores on an intelligence test. It might show the relative frequency with which certain salaries are earned.

The normal curve (Figure 19.13) is used often as a *theoretical* frequency distribution. However, the usual problem is to observe actual data and to describe it pictorially with a graph. Let's do an example. Suppose a city budget lists 100 employees by job category as follows.

Job classification	0	1	2	3	4	5	6	7	8	9	10
Frequency	12	28	18	15	7	10	0	8	0	2	0

Show the graph of the frequency distribution of employees by job category. The graph appears in Figure 19.28. The points are connected by several short straight lines. Graphs of broken lines like this one are called "polygons" (which means "many-sided figure composed of straight lines"). Can you see how the frequency polygon in Figure 19.28 shows relative frequency? It gives a picture of the city's employee situation.

Ordinates of frequency distributions are never negative since frequency counts cannot be negative. Abscissas can be any value, but are usually determined quite practically from the nature of the data being studied. For example, a psychologist studying intelligence might construct a baseline running from 70 to 130. Values below 70 and above 130 might be so rare that they do not need to be indicated on the graph.

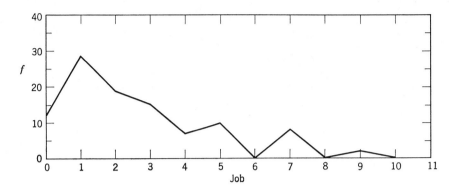

Figure 19.28

TREND LINES

Statistical data concerning relationships generally do not occur in smooth lines, just as the frequency polygon in the previous section was not a smooth curve. The statistician asks such questions as, "When I plot the number pairs, do they seem to occur in a linear or a curved pattern?" Or perhaps he will ask, "What line might be sketched through the data plots that seems to fit the data better than other lines?"

Here are some sets of data (Table 19.3). The first set is plotted in Figure 19.29 and it shows a generally linear pattern. The second set, plotted in Figure 19.30, shows a distinctly curved pattern.

Table 19.3

Set 1	x	1	2	2	3	3	4	4	5	5
	y	1	2	1	3	2	3	4	5	4
Set 2	s	1	2	3	4	3	4	2	1	1	0
	t	0	1	1	2	3	3	4	5	6	7

Examine Figure 19.29 and decide what line seems to fit best. The best line looks like $y = x - .5$. You might sketch this line and see how closely it fits the data points. The points in Figure 19.30 seem to describe a fairly complex curve, and we might be content not to find an equation for this curve. After all, the picture shows the relationship between t and s quite clearly without a formula.

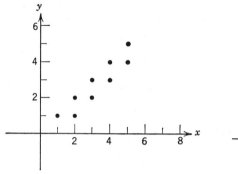

Figure 19.29   Figure 19.30

PROBLEMS

Draw frequency polygons and trend lines for the following sets of data. Label the graph in terms of the variables of interest.

19.13. The frequency distribution of socioeconomic status in an urban school (L = lower, M = middle, U = upper).

Frequency	75	125	50	20	25	15	40	25	0
Classification	0	1	2	3	4	5	6	7	8
	LL	ML	UL	LM	MM	UM	LU	MU	UU

19.14. The distribution of scores obtained on a classroom test.

Frequency	0	2	2	5	8	11	3	2	0
Scores	0	5	10	15	20	25	30	35	40

19.15. The relationship between socioeconomic status and average educational level (in years).

Education	8.1	9.3	9.6	11.0	12.2	14.1	15.2	14.1	14.6
Socioeconomic	0	1	2	3	4	5	6	7	8

19.16. The relationship between age (in years) and average height (in inches).

Age	9	10	11	12	13	14	15	16
Height	50	52	55	60	66	68	69	70

Multivariable Graphing

Most problems in the behavioral sciences involve many variables. Some multivariable (or multivariate) problems can be clarified by graphs, but when graphs require three or more dimensions, it is difficult to draw them on a flat piece of paper. Sometimes three-dimensional models can be constructed for three-variable graphs. Some three-dimensional pictures can be drawn using perspective drawings. A common procedure is to use a contour-line technique, just as the map constructor does, to get altitude, latitude, and longitude all on a two-dimensional graph. Both of these techniques will be illustrated.

Figure 19.31 shows a standard three-dimensional graph. The vector $(4, 2, 7)$ is plotted along with guidelines to show the perspective. The dotted lines, which are not really necessary, show the negative directions of the three axes.

The equation "$Z = X + Y$" can be graphed in a perspective drawing; however, it is difficult to draw it and difficult to understand the resulting drawing. A more convenient way to graph $Z = X + Y$ is by a sort of contour map. We can assign various values to Z and plot in two dimensions an entire family of lines of the general form $X + Y = Z$. Such lines for $Z = -2$,

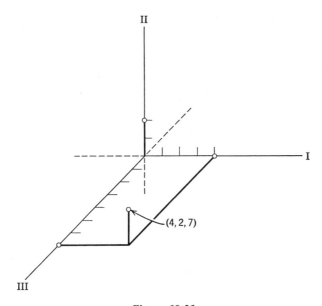

Figure 19.31

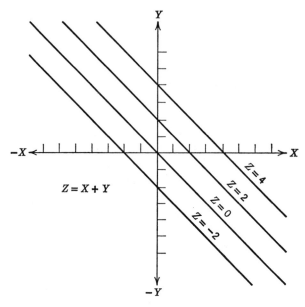

Figure 19.32

0, 2, and 4 appear in Figure 19.32. You can get a good understanding of the three-dimensionality of the system by visualizing the $Z = -2$ line as being two units below the page, the $Z = 0$ line as being exactly on the page, the $Z = 2$ line as being two units above the page, and the $Z = 4$ line as being four units above the page. The "actual" three-dimensional graph would be a plane passing through these four lines as you imagine them as being below, on, or above the page. The lines on the two-dimensional (X, Y) graph are called "projections." The line $X + Y = 4$ is called the "projection of the plane $Z = X + Y$ on the (X, Y) plane for all points such that $Z = 4$ (or $X + Y = 4$)."

Table 19.4

y							
7					1	1	1
6				1	5	5	1
5			1	5	10	5	1
4		1	5	10	5	1	
3	1	5	10	5	1		
2	1	5	5	1			
1	1	1	1				
	1	2	3	4	5	6	7 x

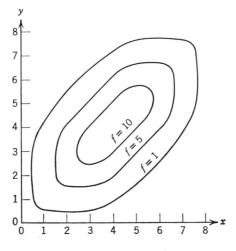

Figure 19.33

This latter approach to three-variable graphs is commonly used in statistics to show joint frequency distributions of two variables and trend lines for predicting a variable from two other variables.

Let's look at an example of a two-variable frequency distribution. Let's call the two variables x and y. Table 19.4 gives f—the *frequency* of each possible (x, y) pair. The contour graph of this frequency distribution appears in Figure 19.33. Notice how this is exactly like a hill plotted on a contour map.

Other Coordinate Systems

INTRODUCTION

The methods used to describe the position of a point in space have been shown to be relatively arbitrary. We choose a reference point and an axis system subjectively, although common sense usually dictates some of the choices. Many other systems are used and others could be devised. Two other types of coordinate systems will be discussed briefly, since they are used occasionally in applied statistics.

POLAR COORDINATES

A polar coordinate system describes the location of a point in space by its *distance* and *direction* from a reference point and reference axis. This is

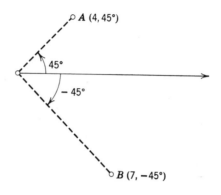

Figure 19.34

similar to how we travel on the earth's surface using a map or compass. We might say, "I am one mile due east of home." This means that "home" is our reference point and magnetic or true north is our reference axis. Our location is one mile from the reference point in a direction 90° clockwise from the reference axis.

A graph is usually drawn with the reference axis pointing to the right. Figure 19.34 shows two points plotted in polar coordinates. The coordinates of point *A* are (4, 45°) or (45°, 4), which means, "Go four units in a direction of 45° counterclockwise." Point *B* is (7, −45°). The Cartesian coordinates of these points would have been about (2.8, 2.8) for *A* and about (5, 5) for *B*.

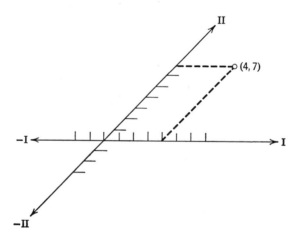

Figure 19.35

OBLIQUE COORDINATES

Oblique coordinate systems are quite similar to Cartesian coordinates, except that the angle between the axes is not 90°. The angle can be any value, but it must be known. The coordinates (a, b), just as is the case with rectangular coordinates, will refer to a point located a-units in the direction of axis one, and b-units in the direction of axis two. Figure 19.35 shows a set of oblique axes, the angle between them being 45°. The coordinates of the point plotted in Figure 19.35 are $(4, 7)$.

Problems for Review

1. Graph the following functions using a Cartesian coordinate system.

a. $3x + 2y + 3 = 0$
b. $6u + 3v - 9 = 0$
c. $u^2 + 2u + 1 = v$

2. Use Cartesian graphs to find the simultaneous solutions to these pairs of equations and inequalities.

a. $3a + b = 4$ and $3a - b = 4$
b. $2x - 2 < y < 2x + 2$
c. $x = y + 2$
$\quad x^2 - 4 = y$

d. Use the same graphs as in 2c, but show the solution set to be

$$x^2 - 4 \leq y \quad \text{and} \quad x > y + 2$$

3. Show graphically that there is no point that satisfies all of these equations simultaneously.

a. $y = 2x + 1$
b. $y = (\frac{1}{3})x + 1$
c. $y = -2x + 3$

4. Plot the following sets of data, and *sketch* the polygons or best-fitting trend lines and curves.

a.

Frequency (%)	15	30	40	10	5
Salary	$3,000	$5,000	$7,000	$9,000	$11,000

b.

Calculating speed	0	19	28	30	27	22	20	18
Age	5	10	15	20	25	30	35	40

c.

Frequency	25	75	115	40	20	10
Reading speed	100	200	300	400	500	600

d.

Comprehension (%)	15	60	75	77	80	80
Reading speed	100	200	300	400	500	600

20

Logarithms

Introduction

NOT too long ago, the logarithm concept and methodology was taught because it was a necessary calculation aid. Today there is seldom a need to use logarithms to speed up calculations, since we have mechanical calculators and electronic calculators and computers.

However, logarithms occur naturally in mathematical formulations independent of their use as computational aids. You will eventually learn to use statistical formulas that contain logarithm functions. There also might be rare occasion to use logarithms as a calculation aid.

For these reasons, we will go into a discussion of logarithm concepts and show how logs and antilogs are determined by tables, how logs are used in multiplication and division, and show special statistical applications of logs. The distinction between common logarithms and natural logarithms will be discussed, since, more often than not, natural logarithms will be more useful to you than common logarithms.

Basic Definitions

GENERAL DEFINITION OF A LOGARITHM

A logarithm is an *exponent* used in a special way. We know that any real number (or a close approximation) can be written in decimal notation. In addition, any real number greater than zero can be written as some power of any other real number.

For example, 9 can be written as "3^2" and 27 can be written as "3^3." Any numbers between 9 and 27 can be written as "3" with a fractional exponent—

253

an exponent between 2 and 3. The number $\frac{1}{3}$ is written "3^{-1}" and $\frac{1}{9}$ can be written "3^{-2}."

Since many numbers must be written with fractional exponents, tables and graphs are used to give these exponents. In our example, all of the numbers were powers of 3, so we say we were working in a *base-3 system*. A table for a base-3 system would give, for selected numbers, the power to which 3 must be raised to give each of the numbers. Here is a small base-3 table.

Number	*Exponent (logarithm)*
81	4
27	3
9	2
3	1
1	0
$\frac{1}{3}$	-1
$\frac{1}{9}$	-2
$\frac{1}{27}$	-3

A full table would show fractional values between these entries.

The word "logarithm" is a synonym for "exponent." However, this synonym is used only in the context in which we are working because "logarithm" is meaningless unless the base is known. We could label the second column as either "exponent" or "logarithm." This language is clarified by the symbols

$$\text{"}\log_3 81 = 4\text{"}$$

which is read, "The log of 81 to the base 3 is 4," or, "The base-3 log of 81 equals 4." "$\log_3 81 = 4$" means, literally, "$3^4 = 81$."

The value of logs is apparent from a problem like this: to what power must 3 be raised to obtain 81?" Or, equivalently "Solve $3^x = 81$ for x." We rewrite the question as

$$\text{"}\log_3 81 = x\text{"}$$

Table 20.1

Log	Antilog
4	81
3	27
2	9
1	3
0	1
-1	$\frac{1}{3}$
-2	$\frac{1}{9}$
-3	$\frac{1}{27}$

and look up "81" in our base 3 table. The table gives us $x = 4$ since $3^4 = 81$. The number 81 will be called the *antilog* of 4 (4 is the log). Thus, our base-three table could have looked like Table 20.1.

Moreover, exponential algebra is simplified using log tables. (You might wish to quickly review parts of Chapter 5.) We know that $3^u \cdot 3^v = 3^{u+v}$ and $3^u \div 3^v = 3^{u-v}$ (from Chapter 5). Also, $(3^u)^v = 3^{u \cdot v}$ and $\sqrt[u]{3} = 3^{1/u}$ These "rules" of exponential algebra can readily be used with log tables.

Here is a set of eight problems (Table 20.2) showing the application of logs to exponential algebra. Study them carefully. See how the *addition* of logs is used for *multiplication* and the *subtraction* of logs is used for *division*. Work out the multiplications to be sure that the log procedure is correct.

Table 20.2

Problem	Logs from Table 20.1	Solution	Antilog from Table 20.1
1. 3×9	$\log_3 3 = 1$ $\log_3 9 = 2$	$1 + 2 = 3$	antilog $3 = 27$
2. $(\frac{1}{27}) \cdot 81$	$\log_3 (\frac{1}{27}) = -3$ $\log_3 81 = 4$	$(-3) + 4 = 1$	antilog $1 = 3$
3. $27 \div 9$	$\log_3 27 = 3$ $\log_3 9 = 2$	$3 - 2 = 1$	antilog $1 = 3$
4. $9 \div (\frac{1}{9})$	$\log_3 9 = 2$ $\log_3 (\frac{1}{9}) = -2$	$2 - (-2) = 4$	antilog $4 = 81$
5. 9^2	$\log_3 9 = 2$	$2 \cdot 2 = 4$	antilog $4 = 81$
6. $\sqrt{81}$	$\log_3 81 = 4$	$(\frac{1}{2}) \cdot 4 = 2$	antilog $2 = 9$
7. $\dfrac{81 \cdot 27}{9 \cdot 3}$	$\log_3 81 = 4$ $\log_3 27 = 3$ $\log_3 9 = 2$ $\log_3 3 = 1$	$4 + 3 - 2 - 1 = 4$	antilog $4 = 81$
8. $((\frac{1}{27})\sqrt{81})^2$	$\log_3 (\frac{1}{27}) = -3$ $\log_3 81 = 4$	$2(-3 + (\frac{1}{2})4) = -2$	antilog $-2 = \frac{1}{9}$

Let's continue the discussion of base 3 by graphing. This is an easy, but crude, way to get the fractional values for our table. A graph shows all values in the range that is graphed, but accuracy is quite limited, and this must be kept in mind. The graph (Figure 20.1) shows the relationship of N (the antilog) to $\log_3 N$. It is the graph of the function $y = \log_3 N$.

Figure 20.1 shows the general shape of curves of the form $y = \log x$. Log curves cross the horizontal axis at $+1$, that is, $\log_b 0 = 1$ for any base ($b^0 = 1$ for any number b). The base is given by the abscissa of the point at which the curve passes $\log_b N = 1$ (b is 3 in Figure 20.1). Accuracy is

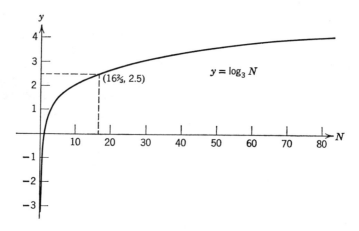

Figure 20.1

extremely bad below $\log_b N = 0$ because the curve approaches zero asymptotically. Frequently, graphs are scaled on a "log scale" that stretches out the horizontal axis, allowing greater accuracy.

Let's use the graph to approximate some arithmetic. Here are some problems.

a. $16\frac{2}{3} \times 3\frac{1}{3} = ?$

b. $43\frac{1}{3} \div 23\frac{1}{3} = ?$

c. $\sqrt{55} = ?$

APPROXIMATE SOLUTIONS

1. a. $\log_3 16\frac{2}{3} + \log_3 3\frac{1}{3}$ is approximately $2.5 + 1.2 = 3.7$. The antilog of 3.7 is about $56\frac{2}{3}$. (The precise product is $55\frac{5}{9}$.)

  b. $\log_3 43\frac{1}{3} - \log_3 23\frac{1}{3}$ is about $3.5 - 2.8$, or $.7$. The antilog of $.7$ is about 2. (The precise answer is $1\frac{6}{7}$.)

  c. $\frac{1}{2}(\log 55)$ is about $(.5)(3.7)$ or 1.85. The approximate antilog of 1.85 is 7.5. To two decimal places, $\sqrt{55}$ is 7.42.

We could do much better with a very large, well-drawn graph, but the principles are illustrated fairly well with even the crude graph.

COMMON AND NATURAL LOGS

It has been stated that log systems can be devised for any base; that is, for any real number all other real positive numbers can be expressed as a power of the base. We have used base 3 as an example. A base of 2 is used occasionally, especially in computer applications. Base 10 is so widely used that logs to base 10 are called "*common logarithms*." The irrational number *e* is also used as a base. Since the number *e* and logs of *e* occur naturally in many mathematical developments, base *e* logs are called "*natural logarithms*."

For review, explain these statements.

a. $\log_{10} 1000 = 3$

b. $\log_e 10 = 2.3026$

Statement *a* means, "$10^3 = 1000$" and statement *b* means, "*e* raised to the power 2.3026 is 10."

Common logs are used so frequently that the base 10 need not be indicated. The symbol "log" *without an indicated base* almost always will mean "log to the *base 10*."

Natural logs are also common enough to have a special symbol. The usual symbol is "ln" ("n" for "natural"). So, statements *a* and *b* can quite correctly be written as

$$\log 1000 = 3 \quad \text{and} \quad \ln 10 = 2.3026$$

PROBLEMS

20.1. Build a log table for the base 2. Show only integral values for the logs from -6 to 7. Use *your* table to do these calculations.

a. $\dfrac{2^4 2^6}{2^3}$ c. $2^{-4} \div 2^{-2}$ e. $2^2 \div 2^7$

b. $\sqrt{2^7 \cdot 2^{-3}}$ d. $(2^{-3})^{-2}$

20.2. Determine x in each of these equations.

a. $\log_3 81 = x$ e. $\log_x 32 = 5$

b. $\log x = 4$ f. $\log 1000 = x$

c. $\log .01 = x$ g. $\log x = -3$

d. $\log_x 10 = 1$ h. $\ln x = 1$

Common Logs

Let's now study thoroughly the common log (base 10) system. A fundamental principle used with logarithms is that any positive real number can be factored into a power of 10 and a number between 1 and 10. For example, study this pattern.

$$7180 = 7.18 \times 10^3$$
$$718 = 7.18 \times 10^2$$
$$71.8 = 7.18 \times 10^1$$
$$7.18 = 7.18 \times 10^0$$
$$.718 = 7.18 \times 10^{-1}$$
$$.0718 = 7.18 \times 10^{-2}$$
$$.00718 = 7.18 \times 10^{-3}$$

Our rule for converting multiplication into addition of logs can now be used to determine the logs for each number in the left-hand column by finding only the number log 7.18. Remember that

$$\log 10^3 = 3$$
$$\log 10^{-3} = -3$$

and, in general, $\log 10^k = k$, for any number k. ("Log 10^k" means, "To what power must 10 be raised to obtain 10^k?" The answer is obvious: 10 must be raised to the kth power to obtain 10^k, so $\log 10^k = k$.) Let's construct a short table of logs and antilogs for the "718" example (Table 20.3).

Table 20.3

Antilog	Log
7180	$\log 7.18 + \log 10^3 = \log 7.18 + 3$
718	$\log 7.18 + \log 10^2 = \log 7.18 + 2$
71.8	$\log 7.18 + \log 10^1 = \log 7.18 + 1$
7.18	$\log 7.18 + \log 10^0 = \log 7.18 + 0$
.718	$\log 7.18 + \log 10^{-1} = \log 7.18 - 1$
.0718	$\log 7.18 + \log 10^{-2} = \log 7.18 - 2$
.00718	$\log 7.18 + \log 10^{-3} = \log 7.18 - 3$

It is apparent that the logs of all numbers in the antilog column are based on the same number—log 7.18. We need to determine only log 7.18 from a common logarithm table to know all of the logs corresponding to the numbers in the antilog column. To four decimal places, log 7.18 = .8561. Let's use this number to complete the "718" table (Table 20.4). The last three entries require further information on standard notation. They could be written accurately as "−.1439," "−1.1439," and "−2.1439," respectively, where these numerals are determined by the indicated subtractions. However, it is more convenient in logarithm handling to keep the

Table 20.4

Antilog	Log
7180	.8561 + 3
718	.8561 + 2
71.8	.8561 + 1
7.18	.8561 + 0
.718	.8561 − 1
.0718	.8561 − 2
.00718	.8561 − 3

numeral ".8561" as part of the log numeral since ".8561" identifies the antilog "7.18." So, logs of numbers between zero and one are usually not written as negative numbers. Here are two common ways of writing negative logs; the second way usually is preferred:

$$\log .718 = \bar{1}.8561 \quad \text{or} \quad 9.8561 - 10$$

$$\log .0718 = \bar{2}.8561 \quad \text{or} \quad 8.8561 - 10$$

$$\log .00718 = \bar{3}.8561 \quad \text{or} \quad 7.8561 - 10$$

So, finally we have out completed "718" log table (Table 20.5).

Table 20.5

Antilog	Log
7180	3.8561
718	2.8561
71.8	1.8561
7.18	0.8561
.718	9.8561 − 10
.0718	8.8561 − 10
.00718	7.8561 − 10

Let's learn two new terms. The ".8561" part of the log numeral is called the "*mantissa*." The mantissa identifies the antilog, but it does not say where the decimal point is to be located. By writing negative logs in the standard notation, we can easily use tables to identify antilogs from logs and logs from antilogs. Most log tables list only mantissas, and mantissas usually are tabled only for antilogs between one and ten. No other values need to be tabled.

The part of the log numeral that identifies the decimal point location is called the "*characteristic.*" Table 20.5, the characteristics are "3," "2," "1," "0," "9 − 10," "8 − 10," and "7 − 10."

The terms "characteristic" and "mantissa" and the concepts just outlined can be clarified by additional examples. Let's consider the numbers 0, 2, and 6. The mantissa of the logs of 0, 2, and 6 are 0, .3010, and .7782, respectively. Let's build a log table for various numbers based on products of 10 and the numbers 0, 2, and 6 (Table 20.6).

Table 20.6

Antilog	Log
1,000	3.0000
600	2.7782
200	2.3010
100	2.0000
60	1.7782
20	1.3010
10	1.0000
6	0.7782
2	0.3010
1	0.0000
.6	9.7782 − 10
.2	9.3010 − 10
.1	9.0000 − 10
.06	8.7782 − 10
.02	8.3010 − 10
.01	8.0000 − 10
.006	7.7782 − 10
.002	7.3010 − 10
.001	7.0000 − 10

Study the patterns. See how the mantissas can always identify whether or not the antilog numeral has a "1," "2," or "6" in it. Also, see how the characteristics identify the location of the decimal point.

Let's do some arithmetic problems using the "0, 2, 6" table. Remember that multiplication is handled by log addition and division is handled by log subtraction. The examples are as follows.

1. $(600 \cdot 20) \div 60$

2. $.6 \div 60$

3. $.02 \times 600 \div 2$

The solutions are as follows.

1. $2.7782 + 1.3010 - 1.7782 = 2.3010$.
   The answer is 200 since log 200 = 2.3010.

2. $(9.7782 - 10) - 1.7782 = 8.0000 - 10$.
   The answer is .01 since log .01 = 8.0000 − 10.

3. $(8.3010 - 10) + (2.7782) - (.3010) = 10.7782 - 10$.
   This is the same as 0.7782, so the answer is 6.

Common Log Tables

We have seen that the characteristic gives the decimal point location in the antilog and the mantissa determines the antilog numeral. Any positive real number can be written as a log. The characteristic will tell the location of the decimal point and the mantissa will tell the numeral. To convert antilogs to logs and logs to antilogs, we need a table that gives only mantissas, since the characteristic is only a decimal point indicator.

Tables for mantissas of common logarithms are in many standard arithmetic and statistics books. It should be apparent that such tables can be quite long, since the log function has an infinite range. The appendix gives an abbreviated common log table. Tabled are mantissa values for numbers between one and 999. The table can be used for almost any number by the choice of the *characteristic*. It is a one-way table whose columns refer to the last digit of the entering number. Here is one row of the table as an example.

	0	1	2	3	4	5	6	7	8	9
24	3802	3820	3838	3856	3874	3892	3909	3927	3945	3962

The tabled number for 241 is .3820. The tabled number for 248 is .3945. Do you see how the columns are used merely to save space? The log of 241 is 2.3820. The log of 24.8 is 1.3945. The log of 2480 is 3.3945. Do you see how logs are obtained from the table?

The accuracy of the table is limited to three significant figures in the antilogs and four significant figures in the logs. Antilogs can be carried to four decimal places by linear interpolation. (Interpolation is discussed in Chapter 21.)

PROBLEMS

Let's learn to find logs and antilogs by practice. Complete the missing values in this table. Save your result for the next section.

	Antilog	*Log*		*Antilog*	*Log*
20.3.	300		20.11.		−2.6160
20.4.	1120		20.12.		4.8621
20.5.	4.83		20.13.		2.9717
20.6.	.641		20.14.		1.9741
20.7.	711		20.15.		4.5065
20.8.	2310		20.16.		−1.8657
20.9.	.0231		20.17.		−2.9085
20.10.	4.12		20.18.		.9956

Problems with Common Log Tables

Let's do more calculations using common logs. We use logs to convert multiplication into addition and division into subtraction. Remember to convert back into antilogs. Estimate your answers to be sure you are relatively close. Use your answers to the preceding set of problems to do these calculations.

20.19. $(1120)(4.83)$

20.20. $(711)(.99)$

20.21. $(.0734)(.0081)$

20.22. $(4.12)(.0413)$

20.23. $32,100 \div 94.2$

20.24. $.0231 \div .0413$

20.25. $(1120)(4.83) \div 711$

20.26. $(4.12)^4$

20.27. $\sqrt{1120}$

20.28. $\sqrt{.0734}$

20.29. $93.7^2 + 94.2^2$ (*Hint.* Add after finding squares.)

20.30. $\sqrt{(.641)(4.83)}$

20.31. $[(.0724)(32,100)]^2$

20.32. $(.0734)(.0081)(.99)$

Natural Log Tables

Natural logs appear frequently in statistical formulas and for this reason are of special interest. We can frequently avoid natural logs by always converting natural logs into common logs, and vice versa.

The conversion is

$$\ln N = 2.3026 \log N$$

or

$$\log N = (2.3026)^{-1} \ln N$$

The constant 2.3026 is ln 10 taken to four decimal places.

Natural logs can be used for arithmetical calculations just as common logs are used. But the problem of the irrational base makes this a little difficult since the characteristic is determined by powers of e, rather than 10. It is not easy, without a lot of experience with e, to decide upon the characteristic of a number like 42.31. How many times must 2.7183 be powered to get 42.31?

So, the importance of the natural log table is not so much for calculations as for merely looking up values and using these in statistical formulas. The problems in this section are all based on standard statistical formulas involving natural logs.

A natural log table appears in the appendix. Natural log tables are not as easily found as common log tables. You can usually find one in calculus or mathematical statistics books.

Problems for Review

Use the natural log table to answer these questions.

1. Evaluate these formulas by making the indicated substitutions.

a. $\chi^2 = -2 \ln \lambda$; $\lambda = 1.23$

b. $\chi^2 = -2 \ln \lambda$; $\lambda = 2.94$

c. $\chi^2 = \ln p_1 + \ln p_2 + \ln p_3$; where $p_1 = .8$, $p_2 = .6$, and $p_3 = .5$

d. $\chi^2 = \sum_{i=1}^{5} p_i$; where $p_1 = p_2 = .4$; $p_3 = p_4 = .6$; and $p_5 = .8$

e. $Z = \frac{1}{2}[\ln (1 + r) - \ln (1 - r)]$; where $r = .6$

f. $Z = \frac{1}{2}[\ln (1 + r) - \ln (1 - r)]$; where $r = .9$

2. In the above problems, "$\chi^2 = -2 \ln \lambda$" can be rewritten as "χ^2 is the natural log of λ^{-2}." Rewrite d and e in a similar kind of sentence showing what operations are implied by the formulas.

3. The equation of the normal curve is

$$y = \frac{1}{\sqrt{2\pi}} e^{-x^2/2}$$

Write an equation for $\ln y$. (*Hint.* $\ln e^u = u$—u is the power to which e must be raised to get e^u.)

4. Let

$$B = 3 \ln (S_1 + S_2 + S_3) - [\ln S_1 + \ln S_2 + \ln S_3]$$

a. Write B without log notation. That is, complete the sentence

"B is the natural log of . . ."

without using logs to indicate exactly what calculations are indicated by the log formula.

b. Find B if $S_1 = 10$, $S_2 = 20$, and $S_3 = 30$.

Tables and Table Reading

Introduction

RESEARCH investigators use statistical tables constantly. In addition, a major research task is often the construction of tables. Most students can use statistical tables with little difficulty; however, occasionally a student or two will have trouble interpreting a table. Many students have trouble interpolating and extrapolating tables.

A statistical table is essentially a number list that defines a functional relationship. A one-way table defines a one-to-one correspondence that could be used in constructing a graph of the function underlying the table. A two-way table defines a two-to-one correspondence. It is used to table functions of two variables. If additional variables need to be included in the function being tabulated, the table usually is broken up into several one- or two-way tables.

Most common statistical functions appear in tables. The purpose of the tables is to present values of the function of interest that can be located and used quickly and accurately.

A major problem in tables arises from the fact that they usually deal with variables that have an infinite number of elements in their range. Since this is true, tables can show only selected values. A table of squares, for example, can show the squares of integers from 1 to 1000. It could show squares of integers from 1 to 10,000 but only show every fifth integer. Frequently, a table of squares will show squares of only the numbers between 1 and 10, but will show them for every number in increments of .01, that is, 1.00, 1.01, 1.02, . . . , 9.99, 10.00. Obviously, the accuracy and utility of tables depends on such things as the increment between entries, the number of entries, and the number of decimal places to which entries are carried.

The standard techniques for determining values that do not appear in a table are *interpolation* and *extrapolation*. Interpolation is the process of estimating a number that is between two tabled numbers. Extrapolation is the process of estimating numbers beyond the range of tabled entries. Usually, extrapolation is quite difficult to do accurately and is to be avoided. Interpolation, on the other hand, can be highly accurate and useful.

In this chapter some examples of one-way and two-way tables will be used. Interpolation in both one- and two-way tables will be explained, and extrapolation will also be discussed.

One-Way Tables

A one-way table can be constructed to show the ordered number pairs relating one variable to another variable by any functional relationship. We have already used one-way tables for calculating squares and square roots and for logarithms. A square root table shows number pairs consisting of a number and its square root. The log table shows number pairs consisting of a log and its antilog.

Many statistical tables show the theoretical frequency distribution of some random variables. The most important example of a one-way frequency distribution table is the normal curve table. This appears in almost every statistics book.

Not all normal curve tables contain the same information, so you should read the description of the table carefully before you use it. Usually the information tabled is, for any number c, the probability of obtaining such a number *or a smaller number*. That is, it tables c and $P\{z \le c\}$ for selected numbers c, where c are numbers in the range of the variable z. Table 21.1

Table 21.1

c	$P\{z \le c\}$
2.5	.99
2.0	.98
1.5	.93
1.0	.84
.5	.69
.0	.50
−.5	.31
−1.0	.16
−1.5	.07
−2.0	.02
−2.5	.01

is a small table of the normal curve. Probability values corresponding to $c < -2.50$ can be considered as zero and values for $c > 2.50$ can be considered as one.

The table can be read in two ways. One way is to choose c and look up $P\{z \le c\}$. For $c = 2.0$, we find $P\{z \le c\} = .98$. The second way is to choose $P\{z \le c\}$ and look up c. We can ask, "For what c does $P\{z \le c\} = .16$?" The table gives the answer, $c = -1.0$.

A normal curve table can be laid out in a way that makes it look like a two-way table. Such a presentation is intended to save space, and possibly to save the researcher time by cutting down the size of the table that must be examined. For example, our little table takes eleven lines. If we wanted to show values in increments of .25 instead of .5, the table would have 21 lines. If we used increments of .01, the table would take up a great deal of space. Let's construct part of our table in increments of .25, but lay it out so that fractional units appear in columns of a matrix.

c	.00	.25	.50	.75
2	.98	.99	.99	1.00
1	.84	.89	.93	.96
0	.50	.60	.69	.77

To use this table, we need to read it as a two-way table, when in fact, it is merely a one-way table cast into a matrix to preserve space. What is $P\{z \le c\}$ if $c = 1.25$? The integral part of "c" instructs you to use the "1" row. The fractional part of "c" instructs you to use the ".25" column. $P\{z \le c\}$ is .89. This technique for condensing one-way tables was used in the log tables of the preceding chapter.

PROBLEMS

21.1. Complete the following table.

	c	$P\{z \le c\}$
a.	2.0	
b.	−1.5	
c.	1.75	
d.	−2.0	
e.		.93
f.		.31
g.		.60
h.		.77

21.2. What is $P\{1.0 \leq z \leq 2.5\}$? (*Hint.* Use $c = 1.0$ and $c = 2.5$.)

21.3. What is $P\{-2.5 \leq z \leq 0\}$?

21.4. Construct a table showing y and $y^2 + 2y + 1$ for values of y from 0 to 4 in increments of .5.

Interpolation in One-Way Tables

Suppose we wished to know $P\{z \leq 1.35\}$ using a normal table. We could look it up in a better table, of course, but suppose we do not have one and must use the little table presented in matrix form in the last section. It does not tell a value of P for $c = 1.35$, but it does show P for $c = 1.25$ and $c = 1.50$. Using these values, we know that

$$.89 < P\{z \leq 1.35\} < .93$$

The standard technique for determining a good estimate between .89 and .93 is *linear interpolation*. We assume that the relationship between P and z is *linear*. Of course, we know that this is not the case because P is an S-shaped curve when plotted (try it). But *in the short range from 1.25 to 1.50*, a straight line might fit the plots quite well. We thus assume linearity in the very short line of interest. There are several ways to use this "local" linearity assumption. For pedagogic reasons, let's graph the two points $(c, P\{z \leq c\})$ that we know, namely (1.25, .89) and (1.50, .93). These can be connected with a straight line (the assumption of "local" linearity). The graph can then be read at the point where $c = 1.35$ intersects the line. The ordinate is the desired answer. By using units of length .005, we can get fairly good accuracy. The solution is plotted in Figure 21.1. The answer is approximately $P\{z \leq 1.35\} = .906$. Even better accuracy would be obtained by more careful graphing and using smaller increments on the graph.

The graphic procedure clearly illustrates why this is called "linear interpolation." We can also solve the problem arithmetically. We know two points and wish to determine an intermediate point. Let's write our problem out in this manner:

c	P
1.50	.93
1.35	?
1.25	.89

Notice that 1.35 is .10 units above 1.25 and .15 units below 1.50. That is, it is $\frac{2}{5}$, or .4, of the distance between 1.25 and 1.50. It seems reasonable to believe that our answer, then, is also greater than .89 by about .4 of the

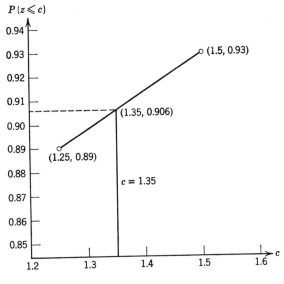

Figure 21.1

distance between .93 and .89. We calculate .93 − .89 = .04 and (.4)(.04) = .016. The interpolated answer is .89 + .016, or .906, which is identical to the result from the graph. The result is the same because the very same principles are involved; in one case we did the work geometrically and in the other case it was done arithmetically. The use of a better table of the normal curve yields $P\{z \le 1.35\} = .9115$. The difference between .9115 and our answer of .906 is one of table accuracy (our table only shows two decimal places) and of nonlinearity. A better table would have allowed the interpolation to be within a much narrower band, thus the "local" linearity assumption would not have led us so far off. In any case, both .9115 and .906 are equal to .91 when rounded to two places.

PROBLEMS

21.5. Use the graph (Figure 21.1) to determine interpolated values of $P\{z \le c\}$ for:

a. $c = 1.42$

b. $c = 1.28$

21.6. Determine arithmetically or geometrically interpolated values for these figures. Partial tables are given.

a. Find by interpolation the square of 95 if $100^2 = 10,000$ and $90^2 = 8100$. What is 95^2 actually? What accounts for the discrepancy?

b. Find Y corresponding to $X = 3.7$ if $Y = 2$ for $X = 3$ and $Y = 5$ for $X = 4$. Evaluate $Y = 3X - 7$ at $X = 3$, 4, and 3.7. Why is there no discrepancy between the interpolated value and the actual value of Y for $X = 3.7$?

c.

d	t
40	2.021
42	?
45	2.014

d.

d	r
300	.113
275	?
200	.138

Two-Way Tables

Two-way tables are matrices listing a three variable function—a two-to-one correspondence between one variable and two other variables. The last section shows how a one-way table can be presented in matrix form giving a two-dimensional appearance. Another type of confusing "matrix-appearing" table is a combination of several related one-way tables shown together in matrix form. It is common, for example, to show several trigonometric functions of angles in one table. This is actually several one-way tables put together to conserve page space and to expedite looking up numbers.

In this section we discuss "real" two-way tables that are based on a two-to-one correspondence between a set of three variables. It is important to make the distinction because of the interpolation problems in a two-way table.

A commonly used two-way statistics table is the F-table. An F-table shows values of a variable called "F" as a function of two constants (parameters)

Table 21.2

d_1	d_2			
	1	3	5	7
2	18.51	19.16	19.30	19.36
4	7.71	6.59	6.26	6.09
6	5.99	4.76	4.39	4.21
8	5.32	4.07	3.69	3.50

called "degrees of freedom." We will call the two parameters d_1 and d_2. Table 21.2 is a partial F-table. The tabulated entries are values of F. To read the table, we must know d_1 and d_2. (These are known in any statistical application of an F-table.) If $d_1 = 6$ and $d_2 = 1$, then $F = 5.99$. If $d_1 = 8$ and $d_2 = 3$, then $F = 4.07$. These can be written $F(6, 1) = 5.99$ and $F(8, 3) = 4.07$. What is $F(2, 5)$ and $F(6, 3)$? They are 19.30 and 4.76, respectively.

Here is another example (Table 21.3). This is a portion of a chi-square (χ^2) table. The tabled numbers are values of a variable called "χ^2." The row names are parameters (degrees of freedom), which we can call "d." The column names are probabilities of obtaining values of χ^2 equal to or less than the table entry. Let χ_t^2 be the table values. Then the column names refer to $P\{\chi^2 \leq \chi_t^2\}$.

Table 21.3

	$P\{\chi^2 \leq \chi_t^2\}$						
d	.90	.80	.70	.50	.30	.20	.10
1	.02	.06	.15	.46	1.07	1.64	2.71
3	.58	1.00	1.42	2.37	3.67	4.64	6.25

So, we can read the table this way: "If $d = 1$, then $P\{\chi^2 \leq .15\} = .70$."

PROBLEMS

21.7. What value of χ^2 corresponds to:

a. $d = 3, P = .50$?

b. $d = 1, P = .80$?

c. $d = 3, P = .20$?

d. $d = 1, P = .10$?

21.8. If $d = 3$, what is the probability of χ^2 being less than or equal to:

a. 4.64?

b. .58?

c. 2.37?

Interpolation in Two-Way Tables

It should be apparent that there are no unique problems in reading tables showing two-way classifications. However, interpolation can be a problem, if interpolation must be made relative to both row and column classifications. Look at the F-table (Table 21.2). If you wanted to know $F(2, 2)$ or $F(3, 3)$, you could estimate these using simple linear interpolation as discussed earlier. But how would you find $F(3, 2)$? This involves interpolating in *two* directions.

Two-way interpolation is handled in three steps. Let's find an estimate of $F(3, 2)$ as an example. First, let's define by two one-way interpolations new row values corresponding to $d_1 = 3$, namely $F(3, 1)$ and $F(3, 3)$. Observe that 3 is halfway between 2 and 4. So, we calculate two new values, each halfway between the numbers above and below them. Our calculations are as follows:

Step 1. Find $F(3, 1)$.

$$F(3, 1) = 7.71 + \frac{18.51 - 7.71}{2}$$

$$= 7.71 + \frac{10.80}{2}$$

$$= 7.71 + 5.40 = 13.11$$

Step 2. Find $F(3, 3)$.

$$F(3, 3) = 6.59 + \frac{19.16 - 6.59}{2}$$

$$= 6.59 + \frac{12.57}{2}$$

$$= 6.59 + 6.285 = 12.875$$

We have obtained interpolated values for $F(3, 1)$ and $F(3, 3)$. We want $F(3, 2)$, so we must interpolate between 13.11 and 12.875.

Step 3. Find $F(3, 2)$ from $F(3, 1)$ and $F(3, 3)$.

$$F(3, 2) = 12.875 + \frac{13.11 - 12.875}{2}$$

$$= 12.875 + \frac{.235}{2}$$

$$= 12.875 + .1175 = 12.9925$$

Our two-way interpolation yields 12.99.

The entire process can be worked by defining first interpolated column entries for $d_2 = 2$. Let's do it again, first finding $F(2, 2)$ and $F(4, 2)$.

Step 1. Find $F(2, 2)$.

$$F(2, 2) = 18.51 + \frac{19.16 - 18.51}{2}$$

$$= 18.835$$

Step 2. Find $F(4, 2)$.

$$F(4, 2) = 6.59 + \frac{7.71 - 6.59}{2}$$

$$= 7.15$$

Step 3. Find $F(3, 2)$ from $F(2, 2)$ and $F(4, 2)$.

$$F(3, 2) = 7.15 + \frac{18.835 - 7.15}{2}$$

$$= 12.9925$$

The result is the same as before.

Let's do an example from the χ^2 table (Table 21.3). What χ_t^2 value corresponds to $d = 2$ and $P = .22$? Let's first find values for $d = 2$, $P = .30$ and $d = 2$, $P = .20$.

Step 1. Find χ_t^2 $(2, .30)$.

$$\chi_t^2(2, .30) = 1.07 + \frac{3.67 - 1.07}{2}$$

$$= 2.37$$

Step 2. Find χ_t^2 $(2, .20)$.

$$\chi_t^2(2, .20) = 1.64 + \frac{4.64 - 1.64}{2}$$

$$= 3.14$$

Step 3. Find χ_t^2 $(2, .22)$.

$$\chi_t^2(2, .22) = 2.37 + (.8)(3.14 - 2.37)$$

$$= 2.986$$

The answer 2.986, can be obtained also by first finding $\chi_t^2(1, .22)$ and $\chi_t^2(3, .22)$. Try it and seen if you can get 2.986.

Problems for Review

1. Find an interpolated value of z for $X = 2$ and $Y = 5$ if

$z = 8$	when	$X = 1$	and	$Y = 2$
$z = 19$	when	$X = 4$	and	$Y = 6$
$z = 16$	when	$X = 1$	and	$Y = 6$
$z = 11$	when	$X = 4$	and	$Y = 2$

2. Estimate χ_t^2 for $d = 2$ and $P = .35$.

22

Computational Accuracy

Introduction

THIS chapter deals with the accuracy to which calculations should be taken when using decimal notation. Frequently, textbooks list fairly rigid formal rules for computational accuracy, a point of view that is probably unnecessary. The accuracy of statistical calculations is usually based on a strange mixture of formal and informal guidelines that are determined by the nature of the problem, the formality of the report that will be written, the complexity of the statistical procedure, and other considerations.

The general guideline is to carry as many decimal places as you reasonably can throughout all calculations. Any rounding that is necessary should be postponed to the final steps, as long as this is practical.

An example of a unique traditional guideline is the correlation coefficient. It is rare to see a correlation coefficient in a published report that does not have exactly two decimal place accuracy. Very rarely there will be three-place accuracy. One-place accuracy or accuracy greater than three places is almost never reported. This is without regard to the accuracy of the original data. Other statistics are also frequently reported in this same standard manner.

However, the formal guidelines for computational accuracy are generally based on the accuracy of the basic data. The general principle is that a computation can be no more reliable than the most unreliable figure upon which it is based.

There is a tendency for persons doing statistical calculations to report results in a way that implies more accuracy than is warranted. For example, you might ask several persons to tell you their income to the nearest $1000. Then if you report that the average income is $5,431.29 you are obviously implying that your measurement of income was highly accurate. In fact, it was quite poor. The average probably should have been reported as "about

274

$5000" or "over $5000" or "between $5000 and $6000." The answer cannot be any more accurate than to the nearest thousand since the data were that imprecise.

This example shows that there is a need for guidelines, if only rough ones. In addition, there are a few technical terms in this chapter that should be a part of your vocabulary.

Significant Figures

The term "*significant figures*" or "*significant digits*" is helpful in discussing calculation accuracy. A decimal system numeral is said to have as many significant figures as there are *nonzero digits plus zeros placed between nonzero digits*. Let's look at some examples.

Numerals with one significant figure:

$$1; 4; 6; 60; 500; 9,000,000; .04; .0006$$

Numerals with two significant digits:

$$12; 34; 1.2; .00021; 7,400; 620,000; .0014$$

Numerals with three significant figures:

$$312; 417; 2.18; .244; .0244; 405; 20,600; .0104$$

Numerals with four significant digits:

$$5416; 412.5; 63,250; 31.24; 50.15; 1,012; 4.001; .001427$$

An exception to this definition is zeros to the right of a decimal point *and* to the right of nonzero digits. Such zeros are counted as significant figures. These numerals all have exactly four significant digits:

$$.1014; .2010; .4000; 500.0; .004610; .004600; .01000$$

This exception is based on the observation that these "extra zeros" are unnecessary *except* to indicate significance. In general, the difference between "5" and "5.0" in statistical data is as follows.

"5" usually means "a number greater than 4.5 but less than 5.5," while "5.0" usually means "a number greater than 4.95 but less than 5.05." That is, "5.0" shows more measurement accuracy than "5."

Zeros to the right of all nonzero digits are ambiguous if there is no decimal point. Conceivably, "6000" could have one, two, three, or four significant figures. You would usually assume that "6000" has only one significant

figure (the "6"). Sometimes dots or bars are used, so that three significant digits might be indicated by "60$\dot{0}$0" or "60$\bar{0}$0." In applied work,
however, this usage is fairly rare, so that "6000" would usually mean "a
number greater than 5500, but less than 6500."

PROBLEMS

How many significant figures are in these numerals?

22.1. 22

22.2. 2.043

22.3. 11.200

22.4. .0010

22.5. .001

22.6. 6,102.4010

22.7. 7.12

22.8. 6.1302

22.9. 6,000,000

22.10. 6,000,500,000

22.11. 1,000,200.34

22.12. .000029

Rounding

"*Rounding*" is the process of discarding unwanted significant figures. The
usual reason for rounding is so that a computational result will have no
greater accuracy implied than is justified. Another reason is to reduce computational difficulty—it is easier to work with few digits in each numeral.

Frequently, the purpose for which the numeral will be used dictates
rounding. For example, suppose a city spends exactly $415,219.26 on city
recreation in a given year. The city auditors need to know the numeral to
the penny. A newspaper reporter might use the numeral "$415,000," lay
persons discussing the budget might say "$400,000," or even "one-half
million." Each person uses the accuracy necessary for his own purposes.

There are standard rules for rounding that should be known. Let's use
the budget figure above as an example.

Significant digits	Numeral
8	415,219.26
7	415,219.30
6	415,219
5	415,220
4	415,200
3	415,000
2	420,000
1	400,000

In lines 6, 4, 3, and 1, the last significant digit retained is the same digit that was used in the preceding line. These were "rounded down" in the sense that the rounding process involved merely dropping nonzero digits. Each numeral in lines 6, 4, 3, and 1 refers to a number *less than* 415,219.26.

The last significant figure retained in lines 7, 5, and 2 has been increased by one over the corresponding digit of the preceding line. These have been "rounded up." The resulting numerals refer to numbers all *larger than* 415,219.26.

If the portion to be discarded is a numeral with a starting digit of "4," "3," "2," "1," or "0," then we round down. If the digit is "6," "7," "8," or "9," we round up. If the digit is "5," we round up if the "5" is followed by any nonzero digit. Check the example using these rules. Here is another example:

$$4,516,293$$

This numeral can be rounded to any of these numerals:

$$4,516,290$$
$$4,516,300$$
$$4,516,000$$
$$4,520,000$$
$$4,500,000$$
$$5,000,000$$

It is rounded to "4,500,000" because the portion discarded, "16,293," refers to a number less than 50,000. It is rounded to "5,000,000" because the portion discarded, "516,293," refers to a number *greater* than 500,000. Therefore, you basically determine whether or not the portion "rounded off" starts with five or more or less to decide whether you round up or down.

Suppose, however, the numeral ends with "5." For example, should "65" be rounded to "60" or "70?" The decision is quite arbitrary; however you should be consistent. A commonly used rule is to round up if the digit preceding the "5" is odd and round down if the digit preceding the "5" is even. Thus, we would have "65" rounded to "60" but "75" rounded to "80." In the long run, you will increase the numerals about one-half the time and decrease them about one-half the time using this rule.

PROBLEMS

Round these numerals to three significant figures.

22.13. 10,211	22.19. 146500
22.14. .02306	22.20. .03210
22.15. 11.55	22.21. 1205
22.16. 6035	22.22. 610.25
22.17. 12,018	22.23. 9018.35
22.18. 146,500.1	22.24. 1.124985

Computational Accuracy

If a and b are two real numbers, then $a + b$, $a - b$, ab, or a/b can have no more accuracy than a or b, *whichever is least accurate*. Also, $\sqrt{a}$ can be no more accurate than a. The word "accuracy" means, basically, the degree to which a number is free of measurement or rounding error.

Most statistical techniques are based on fallible measurement. Perhaps the only truly error-free numbers are frequency counts such as census data. Examples of measures that are likely to be error free include the number of children in a teaching experiment, the number of white rats that turn right in a T-maze, and the number of heads that occur in 20 tosses of a coin—other examples can be constructed, but most, if not all, will deal with frequency counts of one sort or another.

So, most statistical data have measurement error. The error can be small—as in measuring weight with a very good scale—or it can be quite large—as in measuring school achievement with a poorly made classroom examination.

In the case that we have non-error-free numbers, the general rule of accuracy can be interpreted in terms of significant figures—a computational result, in general, deserves no more and no fewer significant figures than *any* number used in the computation. That is, the least accurate number sets the accuracy of the result.

Let's look at some examples. Suppose we wish to add the error-free numbers 16,594 and 12,724, but we will carry only three significant figures in order to simplify the addition. We have

$$
\begin{array}{r}
16,600 \\
+12,700 \\
\hline
29,300
\end{array}
$$

The sum has only three significant figures, so it is usually assumed that the "real sum" is somewhere between 29,250 and 29,350. The real sum is, in fact, 29,318, and this is indeed in the interval.

Suppose that we had chosen to round only one of the numbers. We have

$$
\begin{array}{r}
16,594 \\
+12,700 \\
\hline
29,294
\end{array}
$$

The sum, if it is left as it is, implies that the "real sum" is between 29,293.5 and 29,294.5, which is not true. If the sum is rounded to four digits, (29,290), we might believe that the "real sum" is in the interval 29,285 to 29,295, which is also not true. Only when the sum is rounded to three digits do we get an interval containing the "real sum." The sum, to three significant figures, is 29,300, which is the sum rounded according to the rule. This rounded sum implies that the "real sum" is in the interval 29,250 to 29,350, which is true.

The major error is in reporting a result in a way that implies more accuracy than is justified. The accuracy of the sum is limited to the accuracy of the least accurate addend. Results can be slightly more accurate than the least accurate number, but accuracy *seldom extends beyond one additional significant figure*.

Computational errors in lengthy statistical calculations can be additive. For this reason, rounding should be postponed until a result is reached. Let's suppose we wish to add these numbers, which are based on fallible measurement: 260, 2140, 3040, 70.45, and 82.35. The smallest number of significant figures is two, so we could round each number before adding. We would get 5510, which is rounded to 5500. If we added before rounding, we would get 5592.80 which is rounded to 5600. The "5600" is a more accurate sum that is "5500." The first solution involved a constant bias—each number was rounded down and these rounding errors were accumulated. The second solution involved only one rounding error.

Approximations

It seems appropriate to end this book with a very important suggestion. Rough rounding of all numerals to a few (one or two) places can lead to very quick approximate solutions. Such approximate solutions are helpful in checking more careful calculations for gross errors. This trick is especially helpful in keeping track of decimal points, as was suggested in Part I. You might also use rough approximate answers to solve text examples and problems quickly, although some problems require detailed calculations that can be overlooked with the approximate solutions. Getting approximate square roots is also extremely helpful in keeping decimals straight in square root problems.

PROBLEMS

Here are some problems in which estimates with one digit accuracy can help you be sure that your more accurate solutions are reasonable.

22.25. $\sqrt{42,071.25}$ 22.28. $\sqrt{.00124}$

22.26. $\sqrt{4207.125}$ 22.29. $6.431 \times 50.120 \times .029$

22.27. $\sqrt{.0124}$ 22.30. $\sqrt{7.12^2 + 3.801^2 + 3.92^2}$

Final Word on Rounding

Rules of accuracy are more often violated than not in statistical textbooks and research reports, so don't take these rules too academically. Many statistical text problems lose their value as pedagogic devices if rules of accuracy are followed too strictly. For example, rounding at intermediate steps might lead you to a result slightly different from that in the text, leading you to believe that you have misunderstood a basic principle. Moreover, a few statistical procedures are quite sensitive to rounding error, so frequent rounding is to be avoided in some cases.

Problems for Review

Round these results to reasonable rounded answers.

1. $(.023)(1682) = 38.686$
2. $(.35)(400) = 1400$
3. $439633 \div 299 = 1470.34$
4. $66.19171 \div 7.93 = 8.347$
5. $5.5^2 = 30.25$
6. $\sqrt{6.0} = 2.449$
7. $\sqrt{49} = 7$
8. $5450 + 43 + 161 + 320 = 5974$
9. $108 + 24 + 285 + 147 + 74 + 96 = 734$
10. $(.21)(.304)(.1101) = .007028784$
11. $(1.41)^2 = 1.9881$
12. $4.1023 + .81025 + 541.23 = 546.14255$

Appendix A

Table of Squares, 100—1000

	N^2 0	N^2 1	N^2 2	N^2 3	N^2 4	N^2 5	N^2 6	N^2 7	N^2 8	N^2 9
100	10000	10201	10404	10609	10816	11025	11236	11449	11664	11881
110	12100	12321	12544	12769	12996	13225	13456	13689	13924	14161
120	14400	14641	14884	15129	15376	15625	15876	16129	16384	16641
130	16900	17161	17424	17689	17956	18225	18496	18769	19044	19321
140	19600	19881	20164	20449	20736	21025	21316	21609	21904	22201
150	22500	22801	23104	23409	23716	24025	24336	24649	24964	25281
160	25600	25921	26244	26569	26896	27225	27556	27889	28224	28561
170	28900	29241	29584	29929	30276	30625	30976	31329	31684	32041
180	32400	32761	33124	33489	33856	34225	34596	34969	35344	35721
190	36100	36481	36864	37249	37636	38025	38416	38809	39204	39601
200	40000	40401	40804	41209	41616	42025	42436	42849	43264	43681
210	44100	44521	44944	45369	45796	46225	46656	47089	47524	47961
220	48400	48841	49284	49729	50176	50625	51076	51529	51984	52441
230	52900	53361	53824	54289	54756	55225	55696	56169	56644	57121
240	57600	58081	58564	59049	59536	60025	60516	61009	61504	62001
250	62500	63001	63504	64009	64516	65025	65536	66049	66564	67081
260	67600	68121	68644	69169	69696	70225	70756	71289	71824	72361
270	72900	73441	73984	74529	75076	75625	76176	76729	77284	77841
280	78400	78961	79524	80089	80656	81225	81796	82369	82944	83521
290	84100	84681	85264	85849	86436	87025	87616	88209	88804	89401
300	90000	90601	91204	91809	92416	93025	93636	94249	94864	95481
310	96100	96721	97344	97969	98596	99225	99856	100489	101124	101761
320	102400	103041	103684	104329	104976	105625	106276	106929	107584	108241
330	108900	109561	110224	110889	111556	112225	112896	113569	114244	114921
340	115600	116281	116964	117649	118336	119025	119716	120409	121104	121801
350	122500	123201	123904	124609	125316	126025	126736	127449	128164	128881
360	129600	130321	131044	131769	132496	133225	133956	134689	135424	136161
370	136900	137641	138384	139129	139876	140625	141376	142129	142884	143641
380	144400	145161	145924	146689	147456	148225	148996	149769	150544	151321
390	152100	152881	153664	154449	155236	156025	156816	157609	158404	159201
400	160000	160801	161604	162409	163216	164025	164836	165649	166464	167281
410	168100	168921	169744	170569	171396	172225	173056	173889	174724	175561
420	176400	177241	178084	178929	179776	180625	181476	182329	183184	184041
430	184900	185761	186624	187489	188356	189225	190096	190969	191844	192721
440	193600	194481	195364	196249	197136	198025	198916	199809	200704	201601
450	202500	203401	204304	205209	206116	207025	207936	208849	209764	210681
460	211600	212521	213444	214369	215296	216225	217156	218089	219024	219961
470	220900	221841	222784	223729	224676	225625	226576	227529	228484	229441
480	230400	231361	232324	233289	234256	235225	236196	237169	238144	239121
490	240100	241081	242064	243049	244036	245025	246016	247009	248004	249001
500	250000	251001	252004	253009	254016	255025	256036	257049	258064	259081
510	260100	261121	262144	263169	264196	265225	266256	267289	268324	269361
520	270400	271441	272484	273529	274576	275625	276676	277729	278784	279841
530	280900	281961	283024	284089	285156	286225	287296	288369	289444	290521
540	291600	292681	293764	294849	295936	297025	298116	299209	300304	301401

Source: Herbert Arkin and Raymond R. Colton, *An Outline of Statistical Methods* (4th ed.: New York: Barnes & Noble, 1939).

282

	N^2	N^2	N^2	N^2	N^2	N^2	N^2	N^2	N^2	N^2
	0	1	2	3	4	5	6	7	8	9
550	302500	303601	304704	305809	306916	308025	309136	310249	311364	312481
560	313600	314721	315844	316969	318096	319225	320356	321489	322624	323761
570	324900	326041	327184	328329	329476	330625	331776	332929	334084	335241
580	336400	337561	338724	339889	341056	342225	343396	344569	345744	346921
590	348100	349281	350464	351649	352836	354025	355216	356409	357604	358801
600	360000	361201	362404	363609	364816	366025	367236	368449	369664	370881
610	372100	373321	374544	375769	376996	378225	379456	380689	381924	383161
620	384400	385641	386884	388129	389376	390625	391876	393129	394384	395641
630	396900	398161	399424	400689	401956	403225	404496	405769	407044	408321
640	409600	410881	412164	413449	414736	416025	417316	418609	419904	421201
650	422500	423801	425104	426409	427716	429025	430336	431649	432964	434281
660	435600	436921	438244	439569	440896	442225	443556	444889	446224	447561
670	448900	450241	451584	452929	454276	455625	456976	458329	459684	461041
680	462400	463761	465124	466489	467856	469225	470596	471969	473344	474721
690	476100	477481	478864	480249	481636	483025	484416	485809	487204	488601
700	490000	491401	492804	494209	495616	497025	498436	499849	501264	502681
710	504100	505521	506944	508369	509796	511225	512656	514089	515524	516961
720	518400	519841	521284	522729	524176	525625	527076	528529	529984	531441
730	532900	534361	535824	537289	538756	540225	541696	543169	544644	546121
740	547600	549081	550564	552049	553536	555025	556516	558009	559504	561001
750	562500	564001	565504	567009	568516	570025	571536	573049	574564	576081
760	577600	579121	580644	582169	583696	585225	586756	588289	589824	591361
770	592900	594441	595984	597529	599076	600625	602176	603729	605284	606841
780	608400	609961	611524	613089	614656	616225	617796	619369	620944	622521
790	624100	625681	627264	628849	630436	632025	633616	635209	636804	638401
800	640000	641601	643204	644809	646416	648025	649636	651249	652864	654481
810	656100	657721	659344	660969	662596	664225	665856	667489	669124	670761
820	672400	674041	675684	677329	678976	680625	682276	683929	685584	687241
830	688900	690561	692224	693889	695556	697225	698896	700569	702244	703921
840	705600	707281	708964	710649	712336	714025	715716	717409	719104	720801
850	722500	724201	725904	727609	729316	731025	732736	734449	736164	737881
860	739600	741321	743044	744769	746496	748225	749956	751689	753424	755161
870	756900	758641	760384	762129	763876	765625	767376	769129	770884	772641
880	774400	776161	777924	779689	781456	783225	784996	786769	788544	790321
890	792100	793881	795664	797449	799236	801025	802816	804609	806404	808201
900	810000	811801	813604	815409	817216	819025	820836	822649	824464	826281
910	828100	829921	831744	833569	835396	837225	839056	840889	842724	844561
920	846400	848241	850084	851929	853776	855625	857476	859329	861184	863041
930	864900	866761	868624	870489	872356	874225	876096	877969	879844	881721
940	883600	885481	887364	889249	891136	893025	894916	896809	898704	900601
950	902500	904401	906304	908209	910116	912025	913936	915849	917764	919681
960	921600	923521	925444	927369	929296	931225	933156	935089	937024	938961
970	940900	942841	944784	946729	948676	950625	952576	954529	956484	958441
980	960400	962361	964324	966289	968256	970225	972196	974169	976144	978121
990	980100	982081	984064	986049	988036	990025	992016	994009	996004	998001

Appendix B

Table of Common Logarithms— Five Place

0—50

N	L 0	1	2	3	4	5	6	7	8	9
0	− ∞	00 000	30 103	47 712	60 206	69 897	77 815	84 510	90 309	95 424
1	00 000	04 139	07 918	11 394	14 613	17 609	20 412	23 045	25 527	27 875
2	30 103	32 222	34 242	36 173	38 021	39 794	41 497	43 136	44 716	46 240
3	47 712	49 136	50 515	51 851	53 148	54 407	55 630	56 820	57 978	59 106
4	60 206	61 278	62 325	63 347	64 345	65 321	66 276	67 210	68 124	69 020
5	69 897	70 757	71 600	72 428	73 239	74 036	74 819	75 587	76 343	77 085
6	77 815	78 533	79 239	79 934	80 618	81 291	81 954	82 607	83 251	83 885
7	84 510	85 126	85 733	86 332	86 923	87 506	88 081	88 649	89 209	89 763
8	90 309	90 849	91 381	91 908	92 428	92 942	93 450	93 952	94 448	94 939
9	95 424	95 904	96 379	96 848	97 313	97 772	98 227	98 677	99 123	99 564
10	00 000	00 432	00 860	01 284	01 703	02 119	02 531	02 938	03 342	03 743
11	04 139	04 532	04 922	05 308	05 690	06 070	06 446	06 819	07 188	07 555
12	07 918	08 279	08 636	08 991	09 342	09 691	10 037	10 380	10 721	11 059
13	11 394	11 727	12 057	12 385	12 710	13 033	13 354	13 672	13 988	14 301
14	14 613	14 922	15 229	15 534	15 836	16 137	16 435	16 732	17 026	17 319
15	17 609	17 898	18 184	18 469	18 752	19 033	19 312	19 590	19 866	20 140
16	20 412	20 683	20 952	21 219	21 484	21 748	22 011	22 272	22 531	22 789
17	23 045	23 300	23 553	23 805	24 055	24 304	24 551	24 797	25 042	25 285
18	25 527	25 768	26 007	26 245	26 482	26 717	26 951	27 184	27 416	27 646
19	27 875	28 103	28 330	28 556	28 780	29 003	29 226	29 447	29 667	29 885
20	30 103	30 320	30 535	30 750	30 963	31 175	31 387	31 597	31 806	32 015
21	32 222	32 428	32 634	32 838	33 041	33 244	33 445	33 646	33 846	34 044
22	34 242	34 439	34 635	34 830	35 025	35 218	35 411	35 603	35 793	35 984
23	36 173	36 361	36 549	36 736	36 922	37 107	37 291	37 475	37 658	37 840
24	38 021	38 202	38 382	38 561	38 739	38 917	39 094	39 270	39 445	39 620
25	39 794	39 967	40 140	40 312	40 483	40 654	40 824	40 993	41 162	41 330
26	41 497	41 664	41 830	41 996	42 160	42 325	42 488	42 651	42 813	42 975
27	43 136	43 297	43 457	43 616	43 775	43 933	44 091	44 248	44 404	44 560
28	44 716	44 871	45 025	45 179	45 332	45 484	45 637	45 788	45 939	46 090
29	46 240	46 389	46 538	46 687	46 835	46 982	47 129	47 276	47 422	47 567
30	47 712	47 857	48 001	48 144	48 287	48 430	48 572	48 714	48 855	48 996
31	49 136	49 276	49 415	49 554	49 693	49 831	49 969	50 106	50 243	50 379
32	50 515	50 651	50 786	50 920	51 055	51 188	51 322	51 455	51 587	51 720
33	51 851	51 983	52 114	52 244	52 375	52 504	52 634	52 763	52 892	53 020
34	53 148	53 275	53 403	53 529	53 656	53 782	53 908	54 033	54 158	54 283
35	54 407	54 531	54 654	54 777	54 900	55 023	55 145	55 267	55 388	55 509
36	55 630	55 751	55 871	55 991	56 110	56 229	56 348	56 467	56 585	56 703
37	56 820	56 937	57 054	57 171	57 287	57 403	57 519	57 634	57 749	57 864
38	57 978	58 092	58 206	58 320	58 433	58 546	58 659	58 771	58 883	58 995
39	59 106	59 218	59 329	59 439	59 550	59 660	59 770	59 879	59 988	60 097
40	60 206	60 314	60 423	60 531	60 638	60 746	60 853	60 959	61 066	61 172
41	61 278	61 384	61 490	61 595	61 700	61 805	61 909	62 014	62 118	62 221
42	62 325	62 428	62 531	62 634	62 737	62 839	62 941	63 043	63 144	63 246
43	63 347	63 448	63 548	63 649	63 749	63 849	63 949	64 048	64 147	64 246
44	64 345	64 444	64 542	64 640	64 738	64 836	64 933	65 031	65 128	65 225
45	65 321	65 418	65 514	65 610	65 706	65 801	65 896	65 992	66 087	66 181
46	66 276	66 370	66 464	66 558	66 652	66 745	66 839	66 932	67 025	67 117
47	67 210	67 302	67 394	67 486	67 578	67 669	67 761	67 852	67 943	68 034
48	68 124	68 215	68 305	68 395	68 485	68 574	68 664	68 753	68 842	68 931
49	69 020	69 108	69 197	69 285	69 373	69 461	69 548	69 636	69 723	69 810
50	69 897	69 984	70 070	70 157	70 243	70 329	70 415	70 501	70 586	70 672
N	L 0	1	2	3	4	5	6	7	8	9

Source: Kaj L. Nielsen, *Logarithmic and Trigonometric Tables* (New York: Barnes & Noble,

50—100

N	L0	1	2	3	4	5	6	7	8	9
50	69 897	69 984	70 070	70 157	70 243	70 329	70 415	70 501	70 586	70 672
51	70 757	70 842	70 927	71 012	71 096	71 181	71 265	71 349	71 433	71 517
52	71 600	71 684	71 767	71 850	71 933	72 016	72 099	72 181	72 263	72 346
53	72 428	72 509	72 591	72 673	72 754	72 835	72 916	72 997	73 078	73 159
54	73 239	73 320	73 400	73 480	73 560	73 640	73 719	73 799	73 878	73 957
55	74 036	74 115	74 194	74 273	74 351	74 429	74 507	74 586	74 663	74 741
56	74 819	74 896	74 974	75 051	75 128	75 205	75 282	75 358	75 435	75 511
57	75 587	75 664	75 740	75 815	75 891	75 967	76 042	76 118	76 193	76 268
58	76 343	76 418	76 492	76 567	76 641	76 716	76 790	76 864	76 938	77 012
59	77 085	77 159	77 232	77 305	77 379	77 452	77 525	77 597	77 670	77 743
60	77 815	77 887	77 960	78 032	78 104	78 176	78 247	78 319	78 390	78 462
61	78 533	78 604	78 675	78 746	78 817	78 888	78 958	79 029	79 099	79 169
62	79 239	79 309	79 379	79 449	79 518	79 588	79 657	79 727	79 796	79 865
63	79 934	80 003	80 072	80 140	80 209	80 277	80 346	80 414	80 482	80 550
64	80 618	80 686	80 754	80 821	80 889	80 956	81 023	81 090	81 158	81 224
65	81 291	81 358	81 425	81 491	81 558	81 624	81 690	81 757	81 823	81 889
66	81 954	82 020	82 086	82 151	82 217	82 282	82 347	82 413	82 478	82 543
67	82 607	82 672	82 737	82 802	82 866	82 930	82 995	83 059	83 123	83 187
68	83 251	83 315	83 378	83 442	83 506	83 569	83 632	83 696	83 759	83 822
69	83 885	83 948	84 011	84 073	84 136	84 198	84 261	84 323	84 386	84 448
70	84 510	84 572	84 634	84 696	84 757	84 819	84 880	84 942	85 003	85 065
71	85 126	85 187	85 248	85 309	85 370	85 431	85 491	85 552	85 612	85 673
72	85 733	85 794	85 854	85 914	85 974	86 034	86 094	86 153	86 213	86 273
73	86 332	86 392	86 451	86 510	86 570	86 629	86 688	86 747	86 806	86 864
74	86 923	86 982	87 040	87 099	87 157	87 216	87 274	87 332	87 390	87 448
75	87 506	87 564	87 622	87 679	87 737	87 795	87 852	87 910	87 967	88 024
76	88 081	88 138	88 195	88 252	88 309	88 366	88 423	88 480	88 536	88 593
77	88 649	88 705	88 762	88 818	88 874	88 930	88 986	89 042	89 098	89 154
78	89 209	89 265	89 321	89 376	89 432	89 487	89 542	89 597	89 653	89 708
79	89 763	89 818	89 873	89 927	89 982	90 037	90 091	90 146	90 200	90 255
80	90 309	90 363	90 417	90 472	90 526	90 580	90 634	90 687	90 741	90 795
81	90 849	90 902	90 956	91 009	91 062	91 116	91 169	91 222	91 275	91 328
82	91 381	91 434	91 487	91 540	91 593	91 645	91 698	91 751	91 803	91 855
83	91 908	91 960	92 012	92 065	92 117	92 169	92 221	92 273	92 324	92 376
84	92 428	92 480	92 531	92 583	92 634	92 686	92 737	92 788	92 840	92 891
85	92 942	92 993	93 044	93 095	93 146	93 197	93 247	93 298	93 349	93 399
86	93 450	93 500	93 551	93 601	93 651	93 702	93 752	93 802	93 852	93 902
87	93 952	94 002	94 052	94 101	94 151	94 201	94 250	94 300	94 349	94 399
88	94 448	94 498	94 547	94 596	94 645	94 694	94 743	94 792	94 841	94 890
89	94 939	94 988	95 036	95 085	95 134	95 182	95 231	95 279	95 328	95 376
90	95 424	95 472	95 521	95 569	95 617	95 665	95 713	95 761	95 809	95 856
91	95 904	95 952	95 999	96 047	96 095	96 142	96 190	96 237	96 284	96 332
92	96 379	96 426	96 473	96 520	96 567	96 614	96 661	96 708	96 755	96 802
93	96 848	96 895	96 942	96 988	97 035	97 081	97 128	97 174	97 220	97 267
94	97 313	97 359	97 405	97 451	97 497	97 543	97 589	97 635	97 681	97 727
95	97 772	97 818	97 864	97 909	97 955	98 000	98 046	98 091	98 137	98 182
96	98 227	98 272	98 318	98 363	98 408	98 453	98 498	98 543	98 588	98 632
97	98 677	98 722	98 767	98 811	98 856	98 900	98 945	98 989	99 034	99 078
98	99 123	99 167	99 211	99 255	99 300	99 344	99 388	99 432	99 476	99 520
99	99 564	99 607	99 651	99 695	99 739	99 782	99 826	99 870	99 913	99 957
100	00 000	00 043	00 087	00 130	00 173	00 217	00 260	00 303	00 346	00 389
N	L0	1	2	3	4	5	6	7	8	9

Appendix C

Table of Natural or Naperian Logarithms, .01—11.09

To find the natural logarithm of a number which is 1/10 or 10 times a number whose logarithm is given, subtract from or add to the given logarithm the logarithm of 10.

A 0.00–0.99

-10 should be appended to each logarithm

N	.00	.01	.02	.03	.04	.05	.06	.07	.08	.09
0.0		5.395	6.088	6.493	6.781	7.004	7.187	7.341	7.474	7.592
0.1	7.697	7.793	7.880	7.960	8.034	8.103	8.167	8.228	8.285	8.339
0.2	8.391	8.439	8.486	8.530	8.573	8.614	8.653	8.691	8.727	8.762
0.3	8.796	8.829	8.861	8.891	8.921	8.950	8.978	9.006	9.032	9.058
0.4	9.084	9.108	9.132	9.156	9.179	9.201	9.223	9.245	9.266	9.287
0.5	9.307	9.327	9.346	9.365	9.384	9.402	9.420	9.438	9.455	9.472
0.6	9.489	9.506	9.522	9.538	9.554	9.569	9.584	9.600	9.614	9.629
0.7	9.643	9.658	9.671	9.685	9.699	9.712	9.726	9.739	9.752	9.764
0.8	9.777	9.789	9.802	9.814	9.826	9.837	9.849	9.861	9.872	9.883
0.9	9.895	9.906	9.917	9.927	9.938	9.949	9.959	9.970	9.980	9.990

B 1.00-10.09

N		.00	.01	.02	.03	.04	.05	.06	.07	.08	.09
1.0	0.0	0000	0995	1980	2956	3922	4879	5827	6766	7696	8618
1.1		9531	*0436	*1333	*2222	*3103	*3976	*4842	*5700	*6551	*7395
1.2	0.1	8232	9062	9885	*0701	*1511	*2314	*3111	*3902	*4686	*5464
1.3	0.2	6236	7003	7763	8518	9267	*0010	*0748	*1481	*2208	*2930
1.4	0.3	3647	4359	5066	5767	6464	7156	7844	8526	9204	9878
1.5	0.4	0547	1211	1871	2527	3178	3825	4469	5108	5742	6373
1.6		7000	7623	8243	8858	9470	*0078	*0682	*1282	*1879	*2473
1.7	0.5	3063	3649	4232	4812	5389	5962	6531	7098	7661	8222
1.8		8779	9333	9884	*0432	*0977	*1519	*2058	*2594	*3127	*3658
1.9	0.6	4185	4710	5233	5752	6269	6783	7294	7803	8310	8813
2.0		9315	9813	*0310	*0804	*1295	*1784	*2271	*2755	*3237	*3716
2.1	0.7	4194	4669	5142	5612	6081	6547	7011	7473	7932	8390
2.2		8846	9299	9751	*0200	*0648	*1093	*1536	*1978	*2418	*2855
2.3	0.8	3291	3725	4157	4587	5015	5442	5866	6289	6710	7129
2.4		7547	7963	8377	8789	9200	9609	*0016	*0422	*0826	*1228
2.5	0.9	1629	2028	2426	2822	3216	3609	4001	4391	4779	5166
2.6		5551	5935	6317	6698	7078	7456	7833	8208	8582	8954
2.7		9325	9695	*0063	*0430	*0796	*1160	*1523	*1885	*2245	*2604
2.8	1.0	2962	3318	3674	4028	4380	4732	5082	5431	5779	6126
2.9		6471	6815	7158	7500	7841	8181	8519	8856	9192	9527
3.0		9861	*0194	*0526	*0856	*1186	*1514	*1841	*2168	*2493	*2817
3.1	1.1	3140	3462	3783	4103	4422	4740	5057	5373	5688	6002
3.2		6315	6627	6938	7248	7557	7865	8173	8479	8784	9089
3.3		9392	9695	9996	*0297	*0597	*0896	*1194	*1491	*1788	*2083
3.4	1.2	2378	2671	2964	3256	3547	3837	4127	4415	4703	4990
3.5		5276	5562	5846	6130	6413	6695	6976	7257	7536	7815
3.6		8093	8371	8647	8923	9198	9473	9746	*0019	*0291	*0563
3.7	1.3	0833	1103	1372	1641	1909	2176	2442	2708	2972	3237
3.8		3500	3763	4025	4286	4547	4807	5067	5325	5584	5841
3.9		6098	6354	6609	6864	7118	7372	7624	7877	8128	8379
4.0		8629	8879	9128	9377	9624	9872	*0118	*0364	*0610	*0854
4.1	1.4	1099	1342	1585	1828	2070	2311	2552	2792	3031	3270
4.2		3508	3746	3984	4220	4456	4692	4927	5161	5395	5629
4.3		5862	6094	6326	6557	6787	7018	7247	7476	7705	7933
4.4		8160	8387	8614	8840	9065	9290	9515	9739	9962	0185
4.5	1.5	0408	0630	0851	1072	1293	1513	1732	1951	2170	2388
4.6		2606	2823	3039	3256	3471	3687	3902	4116	4330	4543
4.7		4756	4969	5181	5393	5604	5814	6025	6235	6444	6653
4.8		6862	7070	7277	7485	7691	7898	8104	8309	8515	8719
4.9		8924	9127	9331	9534	9737	9939	*0141	*0342	*0543	*0744
5.0	1.6	0944	1144	1343	1542	1741	1939	2137	2334	2531	2728
5.1		2924	3120	3315	3511	3705	3900	4094	4287	4481	4673
5.2		4866	5058	5250	5441	5632	5823	6013	6203	6393	6582
5.3		6771	6959	7147	7335	7523	7710	7896	8083	8269	8455
5.4		8640	8825	9010	9194	9378	9562	9745	9928	*0111	*0293

Source: Reproduced from C. D. Hodgman, *Mathematical Tables* (7th ed.: Cleveland, Ohio: Chemical Rubber Publishing Co., 1941), pp. 139–142.

B   1.00-10.09 (Concluded)

N	.00	.01	.02	.03	.04	.05	.06	.07	.08	.09
5.5	1.7 0475	0656	0838	1019	1199	1380	1560	1740	1919	2098
5.6	2277	2455	2633	2811	2988	3166	3342	3519	3695	3871
5.7	4047	4222	4397	4572	4746	4920	5094	5267	5440	5613
5.8	5786	5958	6130	6302	6473	6644	6815	6985	7156	7326
5.9	7495	7665	7834	8002	8171	8339	8507	8675	8842	9009
6.0	9176	9342	9509	9675	9840	*0006	*0171	*0336	*0500	*0665
6.1	1.8 0829	0993	1156	1319	1482	1645	1808	1970	2132	2294
6.2	2455	2616	2777	2938	3098	3258	3418	3578	3737	3896
6.3	4055	4214	4372	4530	4688	4845	5003	5160	5317	5473
6.4	5630	5786	5942	6097	6253	6408	6563	6718	6872	7026
6.5	7180	7334	7487	7641	7794	7947	8099	8251	8403	8555
6.6	8707	8858	9010	9160	9311	9462	9612	9762	9912	*0061
6.7	1.9 0211	0360	0509	0658	0806	0954	1102	1250	1398	1545
6.8	1692	1839	1986	2132	2279	2425	2571	2716	2862	3007
6.9	3152	3297	3442	3586	3730	3874	4018	4162	4305	4448
7.0	4591	4734	4876	5019	5161	5303	5445	5586	5727	5869
7.1	6009	6150	6291	6431	6571	6711	6851	6991	7130	7269
7.2	7408	7547	7685	7824	7962	8100	8238	8376	8513	8650
7.3	8787	8924	9061	9198	9334	9470	9606	9742	9877	*0013
7.4	2.0 0148	0283	0418	0553	0687	0821	0956	1089	1223	1357
7.5	1490	1624	1757	1890	2022	2155	2287	2419	2551	2683
7.6	2815	2946	3078	3209	3340	3471	3601	3732	3862	3992
7.7	4122	4252	4381	4511	4640	4769	4898	5027	5156	5284
7.8	5412	5540	5668	5796	5924	6051	6179	6306	6433	6560
7.9	6686	6813	6939	7065	7191	7317	7443	7568	7694	7819
8.0	7944	8069	8194	8318	8443	8567	8691	8815	8939	9063
8.1	9186	9310	9433	9556	9679	9802	9924	*0047	*0169	*0291
8.2	2.1 0413	0535	0657	0779	0900	1021	1142	1263	1384	1505
8.3	1626	1746	1866	1986	2106	2226	2346	2465	2585	2704
8.4	2823	2942	3061	3180	3298	3417	3535	3653	3771	3889
8.5	4007	4124	4242	4359	4476	4593	4710	4827	4943	5060
8.6	5176	5292	5409	5524	5640	5756	5871	5987	6102	6217
8.7	6332	6447	6562	6677	6791	6905	7020	7134	7248	7361
8.8	7475	7589	7702	7816	7929	8042	8155	8267	8380	8493
8.9	8605	8717	8830	8942	9054	9165	9277	9389	9500	9611
9.0	9722	9834	9944	*0055	*0166	*0276	*0387	*0497	*0607	*0717
9.1	2.2 0827	0937	1047	1157	1266	1375	1485	1594	1703	1812
9.2	1920	2029	2138	2246	2354	2462	2570	2678	2786	2894
9.3	3001	3109	3216	3324	3431	3538	3645	3751	3858	3965
9.4	4071	4177	4284	4390	4496	4601	4707	4813	4918	5024
9.5	5129	5234	5339	5444	5549	5654	5759	5863	5968	6072
9.6	6176	6280	6384	6488	6592	6696	6799	6903	7006	7109
9.7	7213	7316	7419	7521	7624	7727	7829	7932	8034	8136
9.8	8238	8340	8442	8544	8646	8747	8849	8950	9051	9152
9.9	9253	9354	9455	9556	9657	9757	9858	9958	*0058	*0158
10.0	2.3 0259	0358	0458	0558	0658	0757	0857	0956	1055	1154

C   10-99

N	0	1	2	3	4	5	6	7	8	9
1	2.30259	39790	48491	56495	63906	70805	77259	83321	89037	94444
2	99573	*04452	*09104	*13549	*17805	*21888	*25810	*29584	*33220	*36730
3	3.40120	43399	46574	49651	52636	55535	58352	61092	63759	66356
4	68888	71357	73767	76120	78419	80666	82864	85015	87120	89182
5	91202	93183	95124	97029	98898	*00733	*02535	*04305	*06044	*07754
6	4.09434	11087	12713	14313	15888	17439	18965	20469	21951	23411
7	24850	26268	27667	29046	30407	31749	33073	34381	35671	36945
8	38203	39445	40672	41884	43082	44265	45435	46591	47734	48864
9	49981	51086	52179	53260	54329	55388	56435	57471	58497	59512

Appendix D

Answers to Problems

CHAPTER 1

2. a, b, d, e, g, h, and j
3. a. 56
   b. 4
   c. 4
   d. 1
   e. 4
   f. 258
   g. 232
   h. 5
   i. 304
   j. 2
   k. 2
   l. 28
   m. 1200
   n. 8

CHAPTER 2

ANSWERS

2.1. a. $\frac{11}{12}$

b. $\frac{7}{8} = \frac{28}{32}$

c. $\frac{39}{63}$

d. $\frac{46}{160}$

e. $\frac{5}{12} = \frac{90}{216}$

f. $\frac{191}{504}$

g. $\frac{12}{18} = \frac{216}{324}$

h. $\frac{30}{84} = \frac{90}{252}$

2.2. a. $14\frac{1}{4}$

b. $14\frac{9}{16}$

c. $17\frac{5}{18}$

d. $17\frac{6}{8}$

e. $45\frac{118}{168}$

2.3. a. $8\frac{3}{36}$

b. $\frac{89}{90}$

c. $12\frac{13}{16}$

d. $\frac{33}{40}$

e. $3\frac{1}{16}$

2.4. a. $\frac{3}{8}$

b. $\frac{12}{56}$

c. 1

d. $\frac{5}{9}$

e. $\frac{21}{480}$

f. $\frac{15}{4} = 3\frac{3}{4}$

g. $\frac{7}{24}$

h. $\frac{1}{2}$

i. $\frac{2}{60}$

j. $\frac{22}{8} = 2\frac{6}{8}$

k. 6

l. $5\frac{1}{2}$

2.5. a. 4

    b. $\frac{5}{12}$

    c. $\frac{7}{8}$

    d. $4\frac{3}{12}$

e. 2

f. $\frac{22}{39}$

g. $3\frac{33}{64}$

h. $2\frac{4}{11}$

Problems for Review

2. a. 12 or 24

    b. 34 or 578

    c. 80

3. a. $\frac{7}{2}$

    b. $\frac{17}{3}$

    c. $\frac{71}{9}$

4. a. $3\frac{1}{7}$

    b. $2\frac{13}{16}$

5. a. $1\frac{5}{8}$

    b. $5\frac{13}{16}$

    c. $11\frac{1}{120}$

    d. $2\frac{5}{12}$

    e. $4\frac{51}{72}$

    f. $3\frac{4}{192}$

    g. $1\frac{1}{3}$

    h. 16

    i. $\frac{3}{11}$

    j. $24\frac{1}{5}$

    k. $328\frac{1}{2}$

d. 90 or 2700

e. 60

f. 720

d. $\frac{517}{23}$

e. $\frac{643}{32}$

f. $\frac{931}{9}$

c. $4\frac{2}{23}$

d. $133\frac{8}{9}$

l. $1\frac{1}{8}$

m. $2\frac{1}{10}$

n. 8

o. $2\frac{32}{33}$

p. 2

q. $\frac{11}{20}$

r. $38\frac{4}{18}$

s. $8\frac{1}{3}$

t. $7\frac{8}{81}$

u. $\frac{111}{113}$

CHAPTER 3

ANSWERS

3.1. 55.2145

3.2. 112.4621

3.3. 653.8

3.4. 52.19

3.5. 90.916

3.6. 37.31

3.7. 13.021

3.8. 938.991

3.9. .79469

3.10. 905.8665

3.11. 108.5

3.12. .18

3.13. .0088

3.14. 10,320

3.15. a. 6.67 (or 6.66 R 6)   d. .008
    b. 4.2   e. 2.625
    c. 6.04   f. 14.375

Problems for Review

1. 26.1624	8. 440.32
2. 267.094	9. 16,419.5
3. 1504.7	10. 41.412
4. 22.36	11. .0048
5. 8.19	12. 7.4547
6. 41.566	13. 20.1
7. 20.465	14. 1.411

15. a. .00625   d. .0000001   g. .3   j. 13.4   m. 7643
    b. .61206   e. 120   h. .05   k. .0000134
    c. 57.753   f. 8.8   i. 71,400   l. .007643

16. 24.67	21. .5635
17. 0	22. 1,040
18. 105.175	23. 22.271
19. .2577	24. 4.18
20. 1.127	25. 217.07

CHAPTER 4

ANSWERS

4.1. a. 4, 5

    b.

Age	Frequency
16	2
17	5
18	10
19	4
20	0
21	3
22	0
23	0
24	1

    c. 0

4.2. $\frac{1}{8}$, .125

4.3. $\frac{4}{10}$, .4

4.4. 8, 8

4.5. $\frac{6}{25}$, .24

4.6.

Age	Proportion
16	.08
17	.20
18	.40
19	.16
20	.00
21	.12
22	.00
23	.00
24	.04
Sum	1.00

4.7.

Age	Percentage
16	8
17	20
18	40
19	16
20	0
21	12
22	0
23	0
24	4
Sum	100%

4.8.

Class One		Class Two	
Grade	*%*	*Grade*	*%*
A	25	A	62.5
B	50	B	25.0
C	20	C	12.5
D	0	D	0.0
F	5	F	0.0
Sum	100%	Sum	100.0%

4.9.

Salary	Frequency	Proportion	Percentage
$24,000	100	.005	.5
23,000	120	.006	.6
22,000	60	.003	.3
21,000	220	.011	1.1
20,000	400	.020	2.0
19,000	100	.005	.5
18,000	0	.000	0.0
17,000	220	.011	1.1
16,000	200	.010	1.0
15,000	580	.029	2.9
14,000	700	.035	3.5
13,000	1280	.064	6.4
12,000	2220	.111	11.1
11,000	3810	.1905	19.05
10,000	1750	.0875	8.75
9,000	2440	.122	12.2
8,000	1700	.085	8.5
7,000	1300	.065	6.5
6,000	840	.042	4.2
5,000	1280	.064	6.4
4,000	680	.034	3.4
Totals	20,000	1.0000	100.00%

Problems for Review

1. a.

Child	Frequency Correct	Proportion	Percentage
Tom	3	.25	25
John	7	.583	58.3
Mary	6	.5	50
Sue	9	.75	75
Larry	7	.583	58.3
Jane	9	.75	75
Bob	3	.25	25
Jack	4	.333	33.3

b.

Item	Frequency Correct	Proportion	Percentage
1.	8	1.00	100.0
2.	6	.75	75
3.	2	.25	25
4.	7	.875	87.5
5.	6	.75	75
6.	0	0.0	0
7.	3	.375	37.5
8.	6	.75	75
9.	4	.5	50
10.	1	.125	12.5
11.	3	.375	37.5
12.	2	.25	25

2.

Symbol	Frequency	Proportion	Percentage
Consonants	47	.522	52.2
Vowels	31	.344	34.4
Numerals	9	.1	10
Punctuation marks	3	.033	3.3
Total	90	.999	99.9%

3. $927,160
4. 4.31%
5. 28.6 hours
6. Males: .495, 49.5%; Females: .505, 50.5%

CHAPTER 5

ANSWERS

5.1. 343
5.2. 625
5.3. 64
5.4. 64
5.5. 49
5.6. 10,000
5.7. 256
5.8. 243
5.9. $\frac{1}{4}$
5.10. 1.5129
5.11. 4
5.12. 11
5.13. 20
5.14. 15
5.15. 11
5.16. .4
5.17. 82369
5.18. 831,744
5.19. 544,644
5.20. 3721
5.21. 5625
5.22. 256
5.23. 904,401
5.24. 272.25
5.25. .00036481

5.26. 3.12
5.27. 5.23
5.28. 9.27
5.29. 2.93
5.34. 50.13
5.35. 371.2
5.37. 6
5.38. 7

5.30. 10.5
5.31. 52.3
5.32. 79.6
5.33. .796
5.36. 8.123

5.39. 24
5.40. 1

Problems for Review

1. a. 9.622404
   b. 1697.44
   c. 27.270901

   d. 121.6609
   e. 1.092727

2. a. 7.38
   b. 2.33

   c. 23.3
   d. 1.016

3. a. 121
   b. 7.81
   c. 4.29

   d. 9.80
   e. 6
   f. 5.10

4. a. 6^7
   b. 5^{12}
   c. 7^7
   d. 1

   e. 25^3 or 5^6
   f. $(7^5)(5)$
   g. 6^{-2} or $1/6^2$
   h. 2^{-2} or $\frac{1}{4}$ or 4^{-1}

CHAPTER 6

ANSWERS

6.1

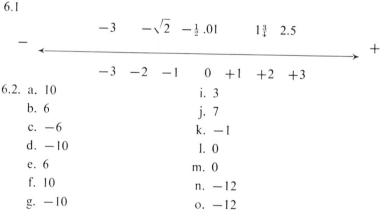

6.2. a. 10
   b. 6
   c. −6
   d. −10
   e. 6
   f. 10
   g. −10
   h. −6

   i. 3
   j. 7
   k. −1
   l. 0
   m. 0
   n. −12
   o. −12
   p. −2

6.3. a. 25
   b. −25
   c. −25
   d. 25

   e. −12
   f. −32
   g. 16
   h. −6

6.4. a. 6
   b. −5
   c. −4
   d. $-\frac{1}{4}$
   e. −40

Problems for Review

1. a. 30
   b. −35
   c. 7
   d. −7

   e. −7
   f. −23
   g. −7
   h. −12

2. a. −8
   b. −8
   c. −4
   d. 19

   e. 16
   f. −8
   g. 5
   h. 12

3. a. 32
   b. −24
   c. −24
   d. 168

   e. 80
   f. 32
   g. 250
   h. 108

4. a. 1.5
   b. −3

   c. $-\frac{1}{18}$
   d. $-16 \div 16 = -1$

5. a. 12
   b. −27
   c. 135
   d. 0

   e. 10
   f. −10
   g. 10
   h. 16

CHAPTER 7

ANSWERS

7.1. "Blacky"
7.2. none needed

7.3. "2" and "$3^2 = 9$"
7.4. "two," "two"

7.5. none needed

7.6. none needed

7.7. "7"

7.8. "one billion"

7.9. none needed

7.10. "VI"

7.11. "VI" and "6"

7.12. a. 42

    b. X

    c. c

    d. $-25Z$

    e. $-25a$

    f. -25

    g. b

    h. 1 $(1 \cdot y = y)$

7.13. $d = (a - 60)^2$,   or $d = (60 - a)^2$

7.14. $k < j + 10$ and $k > j - 10$,   or $j - 10 < k < j + 10$

7.15. $Z = \dfrac{X - 50}{10}$

7.16. $y = \dfrac{x}{10} + 1$,   or $y = \dfrac{x + 10}{10}$

CHAPTER 8

ANSWERS

8.1. a. 17   b. 34   c. 84

8.2. 43

8.3. 22

8.4. 1a. $X_2 = $?   2. $X_4 + X_{10} + X_2 = $?

    b. $X_3 = $?   3. $X_1 - X_6 = $?

    c. $X_1 = $?

8.5. $S_3 = 9$

8.6. $S_j = 0_j$, for $j = 1, \ldots, 25$ (or "for all j")

8.7. $S_{aj} = 0_j + 2E_j$, for $j = 1, \ldots, 25$

8.8. $S_{aj} - S_j = 2E_j$, for $j = 1, \ldots, 25$

8.9. 49

8.10. 49

8.11. 25

8.12. 24

8.13. 28

8.14. 70

8.15. $5 \times 14 = 70$

8.16. 20

8.17. 60

8.18. 140

8.19. 0

8.20. -30

8.21. a. 0
    b. 1
    c. 2
    d. 8
    e. 12

    f. 17
    g. 10
    h. 8
    i. 60
    j. 33

8.22. b. $M = 5$
    c. 28
    d. 28

    e. 37
    f. 244
    g. 1296

8.23. a. $1 + 2 + 3 + 4 + 5 = 15$
    b. $1^2 + 2^2 + 3^2 + 4^2 = 30$
    c. 19
    d. 18 or $(1 \cdot 1 + 1 \cdot 2 + 2 \cdot 1 + 2 \cdot 2 + 3 \cdot 1 + 3 \cdot 2)$
    e. $15 + 5c$ (c is an unknown constant)
    f. $15 + 5j$ (j is merely an unknown constant here)

8.24. a. 168
    b. 84

8.25. 0 (Any factor of 0 will make the product 0.)

8.26. 2

8.27. 120

8.28. 5040

Problems for Review

1. a. 40
    b. 10
    c. 6000

    d. 20,000
    e. 60
    f. 120,000

2. 140 (total number of employees)

3. a. .05
    b. .5

4. $1,110,000 (total annual payroll)

5. $P' = \$1,221,000$ Here are several formulas for P'. Check each one.

   (1) $P' = P + .1P$
   (2) $P' = 1.1P$
   (3) $P' = \sum_{j=0}^{6} n_j(a_j + .1a_j)$
   (4) $P' = 1.1\sum_{j=0}^{6} n_j a_j$

There are several other formulas that are equivalent.

6. 84,000,000
7. a. 120, 160, 420
   c. 4650
   d. $(-20)(0) + (-10)(5) + (10)(0) + (20)(-5) = -150$
   e. (1) 0
      (2) 50

CHAPTER 9

ANSWERS

9.1. -2

9.2. -3 and 3

9.3. $-1, 0, 1, 2, 3, 4$

9.4. all X

9.5. no X

9.6. 1

9.7. 6

9.8. 9

9.9. 60

9.10. -9

9.11. $-a$

9.12. $2 - 3a$

9.13. $3(a + 1)$

9.14. 53

9.15. $b + c + 4$

9.16. $\frac{1}{2}$

9.17. 1

9.18. $\frac{9}{8}$

9.19. $-.6$

9.20. $-.5$

9.21. a. b/a

    b. $6b/a$

    c. abc

    d. ac/b

    e. $3b/7$

9.22. 21

9.23. -7

9.24. 133

9.25. $-7\frac{2}{3}$

9.26. 2

9.27. 3

9.28. 31

9.29. 50

9.30. 5

9.31. all x (identity)

9.32. $-7/38$

9.33. $10/(7a + 16b)$

9.34. 1

9.35. $-20 - 7a$

9.36. $(-15)/(2a + 3b)$

9.37. 0

9.38. $k > 2$ (4, 6, 8)

9.39. $k \leq 2$ ($-8, -6, -4, -2, 0, 2$)

9.40. $k < 1$ $(-8, -6, -4, -2, 0)$

9.41. $k < \frac{3}{2}$ $(-8, -6, -4, -2, 0)$

9.42. $k \leq 8$ (all k)

9.43. $k \leq 8$ (all k)

9.44. 1 and 4

9.45. $-\frac{1}{2}$ and -1

9.46. $-\frac{2}{3}$

9.47. $\sqrt{5}$ and $-\sqrt{5}$

9.48. $-\frac{1}{2}$ and $\frac{1}{3}$

Problems for Review

1. $N \geq 400/d^2$

2. $\sum x^2 = \dfrac{n(n-1)\sigma^2 + (\sum x)^2}{n}$

3. $\bar{X} > 150; \bar{X} < 50$

4. $(\sum X - 5 + 10)/10 = 50.5$

5. $s' = \sqrt{110}$

6. $r' = (r\sqrt{pq})/y$

7. $.25; 0; 0 \leq pq \leq .25$

8. $N = \dfrac{x(1 - C^2)}{C^2}$

9. 30

10. $r = \sqrt{\dfrac{n-1}{n}}$

CHAPTER 10

ANSWERS

10.1. $b = .346$, $b = .594$

10.2. a. $A = -10$, $B = 4$

   b. $X = 1$, $Y = .5$

10.3. $A = 5$, $B = 1$, and $C = -4$

Problems for Review

1. $A_1 = \frac{1}{2}$, $A_2 = -\frac{1}{4}$

2. $X = -10$, $Y = 2$

3. $R = 1$, $S = 2$, $T = 3$

4. $A = 2$, $B = -1$, $C = .5$

CHAPTER 11

ANSWERS

11.1. 3 × 4

11.2. (a) 5   (c) 4
      (b) 6   (d) 7

11.3. a. Smith's sales gross (150,000)
      b. Adams' age (25)
      c. Taylor's data vector (the complete vector)
      d. Sales gross vector (the complete vector)
      e. Total gross sales (650,000)

CHAPTER 12

ANSWERS

12.1. a. $(4, -2, -3)$ d. $(11, 5, 4)$
      b. undefined   e. undefined

      c. $\begin{pmatrix} 6 \\ 1 \\ 1 \end{pmatrix}$ f. $(5, -2, -1, 4)$

12.4. $u_6 + v_6, \ u_i + v_i$

12.5. a. $\begin{pmatrix} 4 \\ 2 \\ 1 \\ 6 \end{pmatrix}$ c. $\begin{pmatrix} 3 \\ 0 \\ 6 \\ 2 \end{pmatrix}$

      b. $(2, -3, 4, 1)$ d. $(6, -1, 5, 7)$

12.6. $v_4, \ v_i$ (No primes are used. Why not?)

12.7. a. $(2, 4, 8, 12)$ e. $(3, 4, 16, 14)$

      b. $(0, -3, 6, -6)$

      c. $(1, 5, -2, 12)$

      d. $(2, 7, 2, 18)$ f. $\begin{pmatrix} 1 \\ -1 \\ 10 \\ 0 \end{pmatrix}$

12.8. 4
12.9. undefined
12.12. 6

12.10. 11
12.11. 4
12.13. 15

Problems for Review

1. a. $t = v + m$
    b. $1'v$ or $v'1$
    c. $v'm$ or $m'v$
    d. $m'm$
    e. $1't$, $t'1$, $1'(v + m)$, or $(v' + m')1$

2.

i	v_i	m_i
1	10	11
2	6	8
3	9	12
4	8	14
5	6	12
6	15	10
7	7	9
8	8	4
9	4	9
10	12	6

a. (21, 14, 21, 22, 18, 25, 16, 12, 13, 18)
b. 180
c. 803
d. (43, 30, 45, 60, 42, 45, 34, 20, 31, 40)
e. 815
f. 983
g. 95

CHAPTER 13

ANSWERS

13.3.

$$3U = \begin{pmatrix} 3 & 12 & 9 \\ 6 & 3 & 18 \\ 3 & 12 & 6 \\ 9 & 3 & 6 \end{pmatrix} \qquad .6U = \begin{pmatrix} .6 & 2.4 & 1.8 \\ 1.2 & .6 & 3.6 \\ .6 & 2.4 & 1.2 \\ 1.8 & .6 & 1.2 \end{pmatrix}$$

13.4. (15, 21)

13.5. (8, 15, 7, 0)

13.6. Undefined product

13.7. Undefined

13.8. $\begin{pmatrix} 5 \\ 21 \\ 9 \\ 5 \end{pmatrix}$

13.9. Undefined

13.10. $\begin{pmatrix} 9 \\ 6 \end{pmatrix}$

13.11

$$B1 = \sum_{j=1}^{m} b_{ij} \qquad \text{There are } n \text{ such sums in } B1 \text{ (i.e., } i = 1, \dots, n\text{).}$$

$$1'B = \sum_{i=1}^{n} b_{ij} \qquad \text{There are } m \text{ such sums in } 1'B \text{ (i.e., } j = 1, \dots, m\text{).}$$

13.12. a. (4 × 2)   g. (5 × 6)
       b. (2 × 2)   h. Undefined
       c. Undefined   i. (4 × 2)
       d. (4 × 4)   j. (4 × 4)
       e. (1 × 5) (row vector)   k. Undefined
       f. (4 × 1) (column vector) l. (2 × 1) (column vector)

13.13. a. $\begin{pmatrix} 9 & 17 & 9 \\ 11 & 17 & 16 \end{pmatrix}$

e. $\begin{pmatrix} -7 & 7 & -14 \\ 7 & 7 & 0 \\ 5 & -1 & 6 \end{pmatrix}$

b. $\begin{pmatrix} 3 & 1 & -2 \\ 8 & 12 & 11 \end{pmatrix}$

f. $\begin{pmatrix} 3 & 8 \\ 13 & 22 \\ 6 & 16 \end{pmatrix}$

c. $\begin{pmatrix} 4 & 8 \\ 8 & 2 \end{pmatrix}$

g. $\begin{pmatrix} 35 & 68 \\ 34 & 78 \end{pmatrix}$

d. $\begin{pmatrix} 4 & 6 \\ 6 & 2 \end{pmatrix}$

h. $\begin{pmatrix} 22 & 14 & 16 \\ 12 & 14 & 10 \end{pmatrix}$

13.14.

$$A' = \begin{pmatrix} 1 & 3 \\ 2 & 1 \\ 3 & 2 \\ 4 & 1 \end{pmatrix} \qquad B' = \begin{pmatrix} 2 & 1 & 4 \\ 1 & 2 & 1 \\ 3 & 1 & 2 \\ 4 & 3 & 5 \end{pmatrix}$$

13.15. (4 × 4), (7 × 7), yes.

13.16. $\sum\limits_{i=1}^{N} x_{2i}x_{3i}$ or $\sum\limits_{i=1}^{N} x_{3i}x_{2i}$

13.17. $\begin{pmatrix} 4 & 2 & 1 \\ 3 & 1 & 2 \\ 1 & 2 & 3 \end{pmatrix} \begin{pmatrix} a \\ b \\ c \end{pmatrix} = \begin{pmatrix} 10 \\ 1 \\ 5 \end{pmatrix}$

$\begin{pmatrix} a \\ b \\ c \end{pmatrix} = \begin{pmatrix} 4 & 2 & 1 \\ 3 & 1 & 2 \\ 1 & 2 & 3 \end{pmatrix}^{-1} \begin{pmatrix} 10 \\ 1 \\ 5 \end{pmatrix}$

13.18. $\begin{pmatrix} 2 & 10 & -2 \\ 0 & -3 & 4 \\ 5 & -2 & 6 \end{pmatrix} \begin{pmatrix} x_1 \\ x_2 \\ x_3 \end{pmatrix} = \begin{pmatrix} 8 \\ 4 \\ 8 \end{pmatrix}$

$\begin{pmatrix} x_1 \\ x_2 \\ x_3 \end{pmatrix} = \begin{pmatrix} 2 & 10 & -2 \\ 0 & -3 & 4 \\ 5 & -2 & 6 \end{pmatrix}^{-1} \begin{pmatrix} 8 \\ 4 \\ 8 \end{pmatrix}$

Problems for Review

1. a. $\begin{pmatrix} 0 & 1 & 0 & -5 \\ -5 & 5 & 0 & 1 \end{pmatrix}$

   g. $\frac{1}{353} \begin{pmatrix} 21 & -5 \\ -5 & 18 \end{pmatrix}$

   b. $\begin{pmatrix} 13 & -6 \\ -6 & 4 \\ 4 & 0 \\ 4 & 8 \end{pmatrix}$

   h. $\frac{1}{22}$

   c. $(9, 4)$

   i. $\begin{pmatrix} 1 & 2 & 4 & 1 \\ 2 & 4 & 8 & 2 \\ 4 & 8 & 16 & 4 \\ 1 & 2 & 4 & 1 \end{pmatrix}$

   d. 9

   j. $\begin{pmatrix} 4 & -2 \\ -2 & 1 \end{pmatrix}$

   e. $\begin{pmatrix} \frac{1}{4} & \frac{1}{4} \\ \frac{3}{8} & -\frac{1}{8} \end{pmatrix}$

   k. $\begin{pmatrix} 18 & 12 \\ 15 & 8 \end{pmatrix}$

   f. $\begin{pmatrix} 18 & 5 \\ 5 & 21 \end{pmatrix}$

   l. $\begin{pmatrix} -3 & -2 \\ -1 & -6 \end{pmatrix}$

2.

   a. $T_s + T_w = \begin{pmatrix} 11 & 11 & 2 & 3 \\ 18 & 11 & 11 & 5 \\ 5 & 9 & 15 & 2 \\ 90 & 18 & 45 & 8 \\ 8 & 3 & 9 & 2 \\ 160 & 40 & 90 & 9 \end{pmatrix}$

b. $1'T_s = (162, 46, 79, 12)$

c. $T_w 1 = \begin{pmatrix} 15 \\ 21 \\ 16 \\ 78 \\ 11 \\ 145 \end{pmatrix}$

d. $1'(T_s + T_w)1$, or $1'T_s 1 + 1'T_w 1 = 585$

e. $T_s'T_s = \begin{pmatrix} 10{,}750 & 2{,}401 & 4{,}706 & 542 \\ 2{,}401 & 578 & 1{,}074 & 132 \\ 4{,}706 & 1{,}074 & 2{,}111 & 245 \\ 542 & 132 & 245 & 32 \end{pmatrix}$

CHAPTER 14

ANSWERS

14.1. a. $\subseteq$ and $\subset$ d. $\nsubseteq$
    b. $\in$ e. $\notin$
    c. $\subseteq$ and $\subset$

1. a. Five is an element of set A.
   b. The set of all even numbers is a proper subset of the set of all numbers.
   c. The set consisting of the numbers 5, 7, and 10 is not a subset of the set consisting of the numbers 5, 7, 20, and 24.
   d. A is the set of counting numbers between one and eight. Or, A is the set of counting numbers 2, 3, 4, 5, 6, and 7.
   e. The complement of D is a proper subset of R.
   f. Seven is not an element of the complement of U.

2. Answers are in the text.

3. a. {all females}
   b. The null set, {h}, {k}, {n}, {h, k}, {h, n}, or {k, n}.
   c. A null set has *no* elements, but A has *one* element, namely the number zero.

d. The set of persons planning to vote Socialist is a null set. The set of persons planning to vote Republican is the complement of the set of persons planning to vote Democratic, and vice versa. There are more elements in the set of Democrat voters than in the set of Republican voters.

e. (1) nine (including the total set of 100)

   (2) 45

   (3) 70

   (4) *One* example.

	M	$\tilde{M}$
C	20	10
$\tilde{C}$	25	45

   (5) A person is in *A* if the person is a male *and* if the person is a college graduate.

   (6) 80

CHAPTER 15

Answers

15.1. a. {1, 3}

   b. {k}

   c. the null set (disjoint sets)

   d. the null set (disjoint sets)

   e. {all x such that $0 \leq x \leq 10$}

15.2. $D \times E = \{(a, \text{male}), (a, \text{female}), (b, \text{male}), (b, \text{female}), (c, \text{male}),$

   $(c, \text{female}), (d, \text{male}), (d, \text{female})\}$

Problems for Review

1. a, d, f, g, and h

3. a. All New Yorkers who are not Republican *and* who are in *either* the upper *or* lower social classes.

   b. All New York Republicans who are in the upper *or* lower social classes.

   c. All middle class New York Republicans.

   d. All New York Republicans (This is $A \cap B$.)

   e. The null set (disjoint sets).

f. All middle class non-Republicans.

g. All persons not from New York.

h. All persons who are *not* simultaneously New Yorkers, Republicans, *and* in the middle class.

i. All New Yorkers *plus* all non-New York middle class Republicans.

j. All Republicans who are *either* New Yorkers *or* middle class *or* both.

CHAPTER 16

ANSWERS

16.1. a. 24
    b. 5040
    c. 40,320
    d. 90
    e. 360
    f. 6
    g. 120

16.2. a. 150
    b. 250
    c. 400
    d. 350
    e. 100
    f. 400

16.3. a. 100
    b. 80
    c. 138
    d. 9
    e. 103
    f. 139

16.4. a. $\binom{25}{20} = \dfrac{25!}{20!\,5!} = 53{,}130$

    b. $\binom{12}{7}\binom{10}{4} = 166{,}320$

    c. $5\binom{21}{2} = 1{,}050$

16.5. a. 21
    b. 3
    c. 10

Problems for Review

1. a. 120
    b. 40,320
    c. 1
    d. 1
    e. 6
    f. 15
    g. 364
    h. 21
    i. 1
    j. 1
    k. 21
    l. 15

2. a. 324
    b. 225
    c. 296
    d. 495
    e. 324
    f. 86
    g. 214
    h. 275

3. a. a, b, c, d, and h
   b. e, f, and g
   c. e, f, and g

5. 990

6. a. 6
   b. 190
   c. 3

CHAPTER 17

ANSWERS

17.1. a. $\frac{26}{52} = \frac{1}{2}$

b. $\frac{12}{52}$

c. $\frac{1}{52}$

17.2. a. $\frac{1}{6}$

b. $\frac{3}{6} = \frac{1}{2}$

c. $\frac{3}{6} = \frac{1}{2}$

17.3. $\frac{1}{6}$

17.4. Look at the horse's racing history. Divide the number of wins by the total number of races.

17.5. a. $\frac{1}{4} + \frac{1}{4} + \frac{1}{4} = \frac{3}{4}$

b. $\frac{1}{4} + \frac{2}{52} = \frac{15}{52}$

c. $\frac{1}{13} + \frac{1}{13} + \frac{1}{13} = \frac{3}{13}$

17.6. a. $\frac{2}{1000} = .002$

b. $\frac{99}{1000} = .099$

c. $\frac{199}{1000} + \frac{100}{1000} = \frac{299}{1000} = .299$

17.7. a. $\frac{1}{16}$

b. $\frac{1}{169}$

c. $\frac{1}{52}$

d. $\frac{1}{13} \cdot \frac{39}{52} = \frac{3}{52}$

e. $\frac{3}{52}$

17.8. a. $\frac{1}{2}$

b. $\frac{16}{52}$

c. $\frac{22}{52}$

d. $\frac{25}{52}$

e. $\frac{32}{52}$

17.9. a. .60

b. .40

c. .50

d. .50

e. .70

f. .20

g. .84

h. .16

i. .36

j. .64

k. .74

17.10. a. .04   d. .85
       b. .02   e. .69
       c. .53

17.11. Number of correct choices   0   1   2   3   4

        Probability $\frac{16}{81}$ $\frac{32}{81}$ $\frac{24}{81}$ $\frac{1}{81}$ $\frac{8}{81}$

Problems for Review

1. a. $(\frac{1}{4})(\frac{12}{51})$

   b. $(\frac{1}{4})(\frac{39}{51}) + (\frac{3}{4})(\frac{13}{51}) = \frac{39}{102}$

   c. $(\frac{1}{4}) + (\frac{12}{52}) - (\frac{3}{52})$

   d. $(\frac{10}{52})(\frac{12}{52}) + (\frac{3}{52})(\frac{11}{51})$

   e. $(\frac{4}{52}) + (\frac{4}{52}) + (\frac{13}{52}) - (\frac{2}{52})$

2. a. .4
   b. .6
   c. .125
   d. $(600 + 400 - 125)/1000$
   e. $\frac{125}{600}$
   f. $\frac{125}{400}$
   g. $(1000 - 600)/1000$ or $1 - (600/1000)$
   h. $\frac{125}{600}$

3. a. .20
   b. .45
   ((a) and (b) are examples of *marginal probability*, the values are obtained from the *margins* of the table when we add down the columns and write the sums on the *margin* as was done in the problem. We can also get *marginal probabilities* for H by adding across the rows.)
   c. .46   f. .071
   d. .77   g. .10
   e. .25   h. .13
                             i. .50

4. a. .02
   b. $(\frac{1}{4})(\frac{8}{20}) = .1$
   c. $(\frac{1}{20})(\frac{8}{20}) = .02$
   d. .02, .01, .02
   e. Method *b* leads to a different probability of inclusion for citizens in each county and thus violates the definition of random sampling (although it *is* a series of completely random operations).

   Method *c* keeps the probability constant for all residents (you can show that each resident in counties 3 and 4 also has a probability of .02 of being included.) This method does allow the possibility of getting a nonrepresentative sample, though.

Method a is the best—it is a simple random sample according to definition. Nonrepresentativeness will occur only by chance using simple random sampling.

5. a. $\frac{1}{4}$
   b. $\frac{1}{16}$
   c. $\frac{7}{80}$
   d. $\frac{1}{20}$
   e. 0
   f. $\frac{1}{4}$
   g. $\frac{7}{40}$
   h. $\frac{1}{20}$

   i.
SE Status	I	II	III	IV	V
Probability	.063	.188	.500	.188	.063

   j.
Achievement	I	II	III	IV
Probability	.8	.2	0	0

   k.
SE Status	I	II	III	IV	V
Cumulative	.063	.250	.750	.938	1.000

6. a. $(\frac{1}{2})^{10}$ or $\frac{1}{1024}$
   b. You should conclude that the sampling was probably not random—for some reason there was bias in the sampling procedure that favored bright children.
   c. $\frac{176}{1024}$

7. a. $\frac{2}{16}$ c. $\frac{8}{16}$
   b. $\frac{6}{16}$ d. $\frac{4}{16}$

8. a. $\binom{13}{5} \div \binom{52}{5}$

   b. $4\binom{13}{5} \div \binom{52}{5}$

   c. $4 \div \binom{52}{5}$

9.
No. Correct	0	1	2	3	4	5	6	7	8
Probability	$\frac{1}{256}$	$\frac{8}{256}$	$\frac{28}{256}$	$\frac{56}{256}$	$\frac{70}{256}$	$\frac{56}{256}$	$\frac{28}{256}$	$\frac{8}{256}$	$\frac{1}{256}$

CHAPTER 18

1. $x = \sqrt{3}$ and $-\sqrt{3}$ (irrational)
2. $x = 1.6$ and $-1.6(\frac{16}{10})$
3. $x = \frac{1}{6}$ and $-\frac{1}{6}(.1\bar{6})$
4. $x = .5$ and $-.5(\frac{1}{2})$
5. $x = -1$

6. $y = \frac{25}{9}(2.\bar{7})$
7. $y = \sqrt{\frac{5}{3}}$ and $-\sqrt{\frac{5}{3}}$ (irrational)
8. $x = 1.\bar{1}(\frac{10}{9})$
9. $x = 4$ and 3
10. irrational

CHAPTER 19

ANSWERS

19.1.

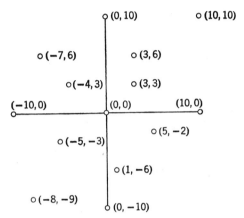

19.2. The variables must be named and equations of the form $y = a + b$ constructed. Call the abscissas A and the ordinates B.

Fig. 19.10. a. $B = A$
  b. $3B = A$ or $A = \frac{1}{3}B$
  c. $B = -2A$ or $A = -\frac{1}{2}B$

Fig. 19.11. a. $B = A + 2$
  b. $B = A$
  c. $B = A - 3$

19.3.

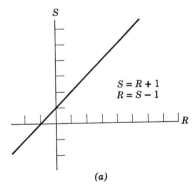

$S = R + 1$
$R = S - 1$

(a)

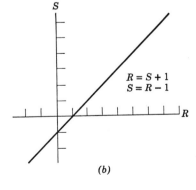

$R = S + 1$
$S = R - 1$

(b)

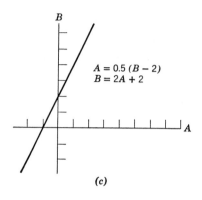

(c)

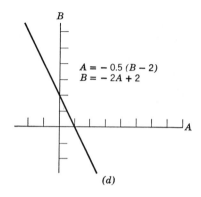

(d)

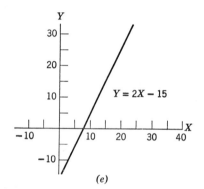

(e)

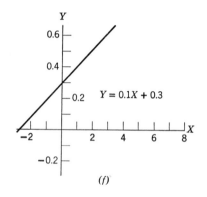

(f)

19.4.

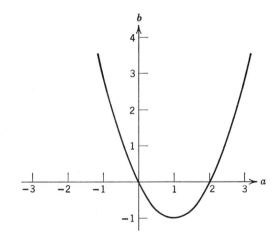

19.5.

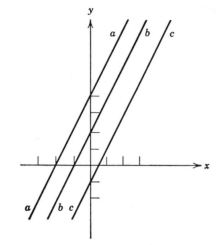

19.6.

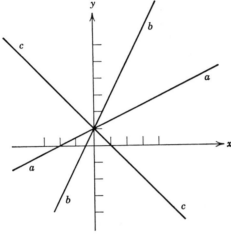

19.7.

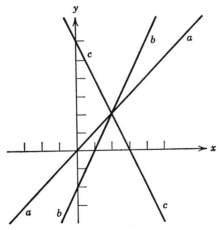

19.8.

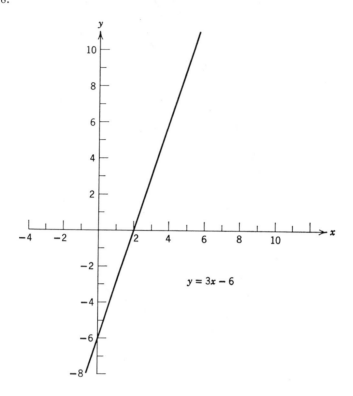

$y = 3x - 6$

19.9.

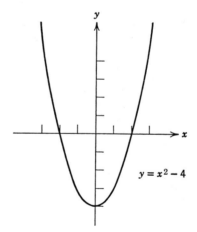

$y = x^2 - 4$

19.10.

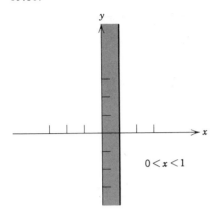

$0 < x < 1$

19.11.

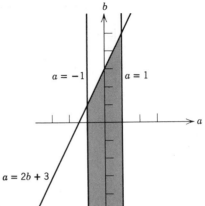

$a = -1$ $a = 1$

$a = 2b + 3$

19.12.

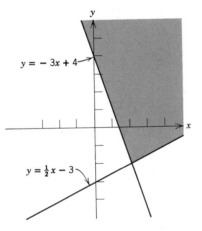

$y = -3x + 4$

$y = \frac{1}{2}x - 3$

19.13.

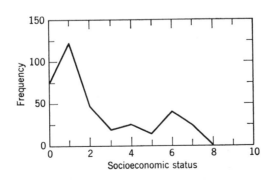

19.14.

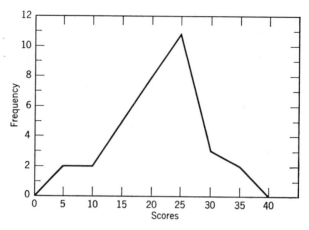

19.15.

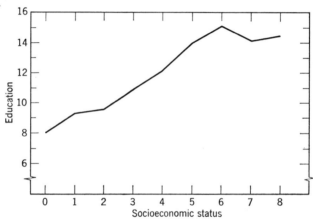

19.16.

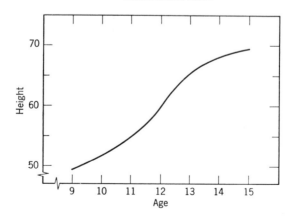

Problems for Review

1a–1b

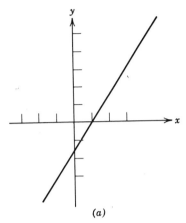

(a)

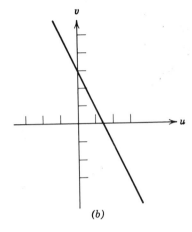

(b)

1c

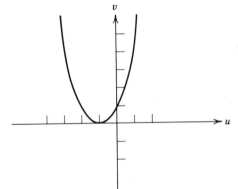

2a–2b

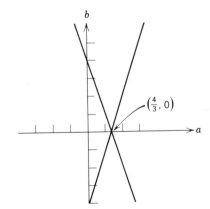

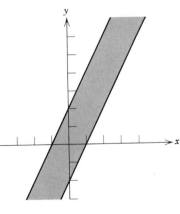

2c–2d

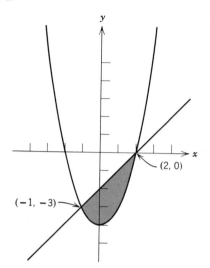

(2, 0)

(−1, −3)

3

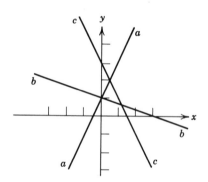

4a, 4b

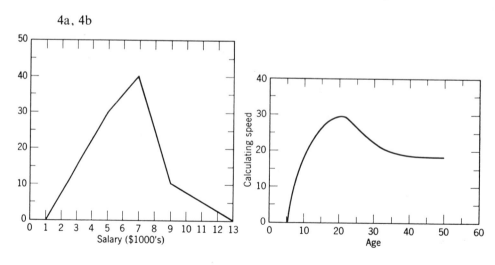

4c, 4d

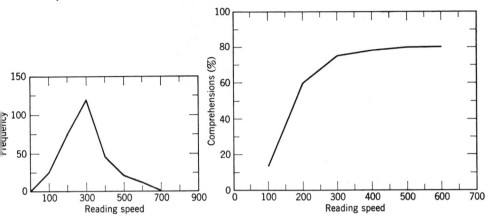

CHAPTER 20

ANSWERS

20.1. *Log*   *Antilog*

Log	Antilog
7	128
6	64
5	32
4	16
3	8
2	4
1	2
0	1
-1	.5
-2	.25
-3	.125
-4	.0625
-5	.03125
-6	.015625

a. 128   d. .64
b. 4   e. .03125
c. .25

20.2. a. 4   e. 2
      b. 10,000 f. 3
      c. -2 g. .001
      d. 10   h. e

20.3. 2.4771	20.11. .0413
20.4. 3.0492	20.12. 72,800
20.5. 0.6839	20.13. 937
20.6. −0.8069	20.14. 94.2
20.7. 2.8519	20.15. 32,100
20.8. 4.3636	20.16. .0734
20.9. −1.3636	20.17. .0081
20.10. 0.6149	20.18. 9.9

(Your answers can differ slightly as a result of rounding errors.)

20.19. 5410	20.26. 288
20.20. 704	20.27. 33.5
20.21. .000595	20.28. .271
20.22. .170	20.29. 17,653
20.23. 341	20.30. 1.76
20.24. .559	20.31. 5,400,000
20.25. 7.61	20.32. .000589

Problems for Review

1. a. $(-2)(.107) = -.214$

   b. -2.157

   c. $(9.777 - 10) + (9.489 - 10) + (9.307 - 10) = -1.427$

   d. -3.077

   e. $\frac{1}{2}(.470 - (-.916)) = .693$

   f. 1.472

2. d. χ^2 is the natural log of $\prod_{i=1}^{5} p_i$ (or $p_1 p_2 p_3 p_4 p_5$)

   e. Z is the natural log of

   $$\sqrt{\frac{1+r}{1-r}}$$

3. $\ln y = -(\frac{1}{2}) \ln 2\pi - x^2/2$ or $\ln y = -(\frac{1}{2})[\ln 2 + \ln \pi + x^2]$

4. a. $B = \dfrac{(S_1 + S_2 + S_3)^3}{S_1 \cdot S_2 \cdot S_3}$

   b. 36

CHAPTER 21

ANSWERS

21.1. a. .98   e. 1.5
      b. .07   f. −.5
      c. .96   g. .25
      d. .02   h. .75

21.2. .15

21.3. .49

21.4.

y	.0	.5
4	25.00	
3	16.00	20.25
2	9.00	12.25
1	4.00	6.25
0	1.00	2.25

21.5. a. .917
      b. .894

21.6. a. Interpolated—9050
         Actual   —9025
         Discrepancy is a result of the nonlinearity of the function $Y = X^2$.

      b. Interpolated—$Y = 4.1$
         Actual   —$Y = 4.1$
         The function $Y = 3X - 7$ is linear.

      c. 2.0182 or 2.018

      d. .11925 or .119

21.7. a. 2.37
      b. .06
      c. 4.64
      d. 2.71

21.8. a. .20
      b. .90
      c. .50

Problems for Review

1. $z = 15$
2. $\chi_t^2 = 2.131$

CHAPTER 22

ANSWERS

22.1. 2	22.7. 3
22.2. 4	22.8. 5
22.3. 5	22.9. 1
22.4. 2	22.10. 5
22.5. 1	22.11. 9
22.6. 8	22.12. 2

22.13. 10,200	22.19. 146,000
22.14. .0231	22.20. .0321
22.15. 11.6	22.21. 1200 (treat 0 as even)
22.16. 6040	22.22. 610
22.17. 12.0	22.23. 9020
22.18. 147,000	22.24. 1.12

22.25. The answer to Problem 22.25 is some number a little higher than 200 since $\sqrt{40,000} = 200$.

22.26. The second answer is near 65 since $\sqrt{4000}$ is close to 65.

22.27. The third answer is about .1 since $\sqrt{.01} = .1$.

22.28. The fourth answer is about .035 since $\sqrt{.0012}$ is between .03 and .04.

22.29. This product is about 9 since $6 \times 50 \times .03 = 300 \times .03 = 9$.

22.30. This answer is near 9 since $7^2 + 4^2 + 4^2$ is about $50 + 32 = 82$ and $9 = \sqrt{81}$.

Problems for Review

1. 39	7. 7.0
2. 1000	8. 6$\bar{0}$00
3. 1470	9. 730
4. 8.35	10. .007$\bar{0}$
5. 3$\bar{0}$	11. 1.99
6. 2.4	12. 546.14

Index

323